PROPERTIES of PETROLEUM RESERVOIR FLUIDS

PROPERTIES of PETROLEUM RESERVOIR FLUIDS

Emil J. Burcik

Professor Emeritus of Petroleum and Natural Gas Engineering
The Pennsylvania State University

IHRDC

International
Human Resources Development
Corporation
8 Arlington Street
Boston, MA 02116 U.S.A.

ISBN: 0-934634-00-9

Library of Congress Catalog Number: 57-5906

This book is dedicated to
M. B.

PREFACE TO IHRDC PUBLICATION

This volume was reprinted by IHRDC because of the strong demand in our reservoir engineering short courses and elsewhere for an excellent book on the fundamentals of reservoir rock and fluid properties. Originally published by John Wiley & Sons, Inc. in 1957 (second printing in 1961), copyright has been reassigned to Burcik and he has given us special permission to reprint this classic book which has been out of print for a number of years. The basis for petroleum conservation and maximization of recovery is to be found in a proper application of reservoir engineering. A systematic exposition of reservoir engineering fundamentals is to be found in a thorough reading of this volume. Many reservoir engineers in the industry today received their introduction to the subject by using this book; we hope that many more in the future will follow their path.

David A. T. Donohue
Publisher

ABOUT THE AUTHOR

Dr. Emil J. Burcik is Professor Emeritus of Petroleum and Natural Gas Engineering at the Pennsylvania State University where he has gained a strong reputation for his teaching of the introductory courses in reservoir engineering. Prior to his current position he was a member of the faculty at the University of Oklahoma and has been visiting professor at both the University of Zulia, Maracaibo, Venezuela and the University of Trondheim, Norway. After receiving his Ph.D. degree from the California Institute of Technology, Dr. Burcik gained industrial experience at Proctor & Gamble Corporation and government experience at the U.S. Bureau of Mines before entering the academic field. He is the author of numerous technical publications including this widely used textbook.

PREFACE

Petroleum engineering practice is constantly changing. This is due not only to the continual introduction of new methods and techniques but to changing economic factors as well. Nevertheless, there are certain fundamental concepts which form the foundation on which the art of petroleum engineering is based. The purpose of this book is to describe these basic concepts in so far as reservoir fluids are concerned.

Reservoir fluids are generally complex mixtures of hydrocarbons existing as liquid-gas systems under high pressures and relatively high temperatures. Since an important aspect of petroleum engineering is predicting the future behavior of a petroleum reservoir when it is produced in some prescribed manner, it is necessary to know the behavior of reservoir fluids as a function of temperature and pressure. Many of the laws which describe the behavior of simple, ideal fluid systems are known, but when dealing with the complex systems existing in petroleum reservoirs, it is necessary to recognize the deviations from ideal behavior. This book describes not only the properties of ideal fluids but also presents empirical correlations which must be employed for a more realistic approach to the problem.

The fundamental concepts which underlie engineering practice have not necessarily been discovered and developed by engineers. Petroleum engineering practice borrows heavily from the basic sciences of mathematics, physics, and chemistry. The basic concepts of reservoir fluid behavior have been developed to a large degree by physical chemists and physicists. Consequently, a great deal of the material presented in this text can properly be considered to be a part of the science of physical chemistry. However, a previous knowledge of physical chemical principles is not necessary for an understanding of the subject matter contained in this text since I have attempted to develop the necessary principles as they are needed.

This book evolved from notes prepared and developed by me over the past six years. This material forms the subject matter for a one-semester petroleum engineering course at The Pennsylvania State University. This course along with one dealing with the fundamental properties of reservoir rocks and rock-fluid systems are prerequisites for more advanced courses in reservoir engineering. These basic courses are an introduction to the fundamentals which must be mastered by the student before the treatment of more complex systems can be attempted.

A list of problems has been included at the end of each chapter in an effort to illustrate more fully the principles that have been discussed. It is my conviction that students will benefit greatly by working these problems since many important concepts are illustrated in their solution. For the benefit of those who are using this book without the aid of formal instruction, answers to some of the problems have been included.

I wish to thank my colleagues and former students who have made this book possible. Their suggestions and criticisms have been an invaluable aid. In particular, I wish to thank Dr. John C. Calhoun and Dr. David T. Oakes for their valuable comments and aid in preparing the manuscript.

EMIL J. BURCIK

University Park, Pennsylvania
November, 1956

CONTENTS

Chapter 1 The Properties of Naturally Occurring Petroleum Deposits 1

2 Behavior of Gases 16

3 Phase Behavior of Liquids 38

4 Qualitative Phase Behavior of Hydrocarbon Systems 48

5 Quantitative Phase Behavior 79

6 Reservoir Fluid Characteristics 101

7 Elementary Applications of Reservoir Fluid Characteristics 155

Appendix A 181

Appendix B 183

Appendix C 187

Index 189

CHAPTER 1

THE PROPERTIES OF NATURALLY OCCURRING PETROLEUM DEPOSITS

Petroleum deposits are naturally occurring mixtures of organic compounds found within the strata of the earth. Those that occur in the gaseous state are generally referred to as natural gas whereas those occurring as liquids are known as petroleum oil or crude oil. Solid deposits are known as tars and asphalts, but, since many of these are actually semisolids or plastic solids, they should be classified as such.

The exact origin of petroleum is not known, but scientific opinion is in favor of origin from the transformation of plant, animal, and marine organisms after marine deposition within finely divided muds. Consensus holds that, after being formed, petroleum was transferred from its source beds to the rock strata where it is now found.

Petroleum is found to occur in local accumulations within the strata of the earth where conditions for accumulation and storage are favorable. This may be within a stratum of rock which has been exposed at the surface of the earth or buried thousands of feet within the earth's interior. Consequently, the pressure and temperature environment under which petroleum exists may vary over wide limits. For this reason, the behavior of petroleum as a function of pressure and temperature is important.

Petroleum deposits will vary widely in chemical composition and those obtained from different localities may have entirely different physical and chemical properties. In spite of this diversity, the bulk of the chemical compounds found in petroleum are hydrocarbons. As the name implies, a hydrocarbon is a compound consisting of carbon and hydrogen only. Even though only two elements are present, the number of compounds of this type is very large. This is due to the fact that the carbon atom has the ability to combine with itself and form long chains. Other elements possess this property to only a limited degree. Thus, oxygen can combine with itself and form O_2 while ozone, a relatively unstable compound, consists of three oxygen atoms. Similarly, sulfur can combine with itself to form the molecule S_8 in the vapor state. However, carbon compounds are known which have carbon chains consisting of hundreds of carbon atoms.

The Paraffin Series. The hydrocarbons may be classed according to structure of the molecule. A very important class is the paraffin hydrocarbons. This series is characterized by the fact that the carbon atoms are arranged in open chains (i.e., not closed rings) and are joined by single bonds. That is, one valence of each carbon is used to form the chemical bond between adjacent carbon atoms in the chain.

The first member of the paraffin series is methane. It consists of a single carbon atom and four hydrogen atoms. Its molecular formula is CH_4. To indicate its structure one writes the structural formula as

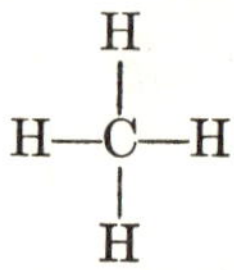

The second member is ethane whose formula is C_2H_6. Its struc-

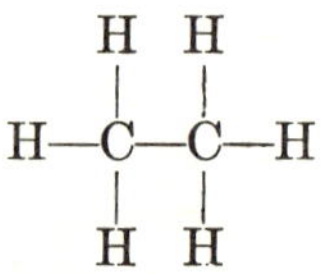 or CH_3CH_3

tural formula is as shown, which indicates that the molecule consists of a chain of two carbon atoms, joined by a single bond, and six hydrogens, three on each carbon as required by the fact that the valence of carbon is four. Some of the succeeding members of this series are listed

Name	Formula	Structural Formula
Propane	C_3H_8	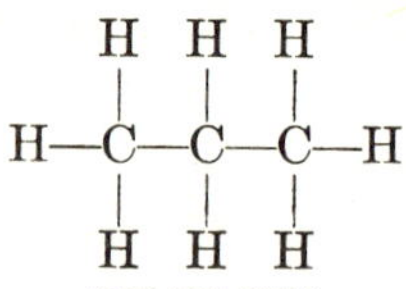 ($CH_3CH_2CH_3$)

Name	Formula	Structural Formula
Butane	C_4H_{10}	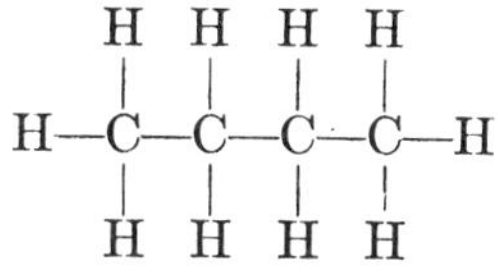 $(CH_3CH_2CH_2CH_3)$
Pentane	C_5H_{12}	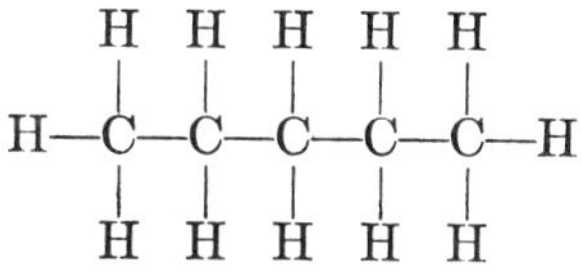 $(CH_3CH_2CH_2CH_2CH_3)$

here. The names of the hydrocarbons in this series all end in -ane. The higher members are named simply by taking the Greek prefix

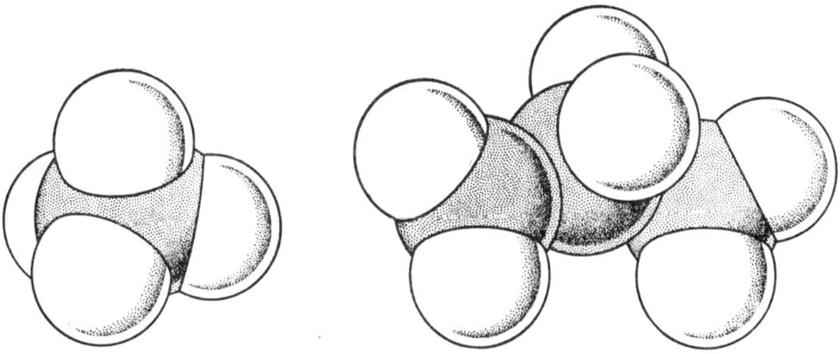

FIG. 1. Models showing the structure of methane and propane.

indicating the number of carbon atoms in the molecule and adding the suffix -ane. Thus, C_6H_{14} is known as hexane, C_7H_{16} is heptane, C_8H_{18} is octane, and so on. The general formula for this series of hydrocarbons is C_nH_{2n+2}.

It should be understood that the structural formulas given above *do not* represent the actual structure of these molecules. The four valences of the carbon atom are directed toward the corners of a regular tetrahedron. Consequently, the hydrogens are not all in one plane and the carbon atoms are not in a straight chain. The structural formula merely represents the fact that a certain number of carbons are joined together by single bonds to form an open chain and the remaining valences form bonds with hydrogen. Models of methane and propane showing the actual structure of these compounds are illustrated in Figure 1.

Isomerism. In the case of methane, ethane, and propane, only one structure can be written that satisfies the requirements of the paraffin class of hydrocarbons. This is not so for butane and the higher members of this series. For example, two structures are possible for butane, both of which have the same molecular formula. Both formulas

Normal Butane Isobutane

satisfy the requirements for a paraffin hydrocarbon. Both are open-chain compounds of hydrogen and carbon only, and the bonds between the carbons are single. However, these compounds have entirely different properties. Isobutane has a boiling point of 10.9° F while normal butane boils at 31.1° F. Other physical and chemical properties of these two compounds are also different. Compounds that have the same molecular formula but have different structures and consequently different physical and chemical properties are called isomers.

Pentane has three isomers whose structural formulas are shown here. Again each of these isomers has the molecular formula C_5H_{12}.

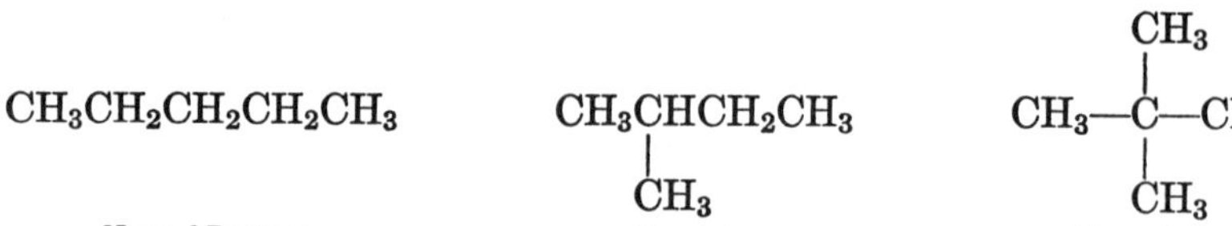

Normal Pentane Isopentane Neopentane

The student should convince himself that there are three, and only three, isomers possible for this hydrocarbon.

The number of possible isomers increases rapidly as the number of carbons in the molecule increases. Hexane has five isomers; heptane has nine. It has been calculated that the hydrocarbon whose formula is $C_{40}H_{82}$ has more than 69×10^{12} isomers.

Nomenclature. The question arises as to how the various isomers of a hydrocarbon can be named and distinguished. When the number of possible isomers is small this offers no problem. The two isomers of butane are named normal butane and isobutane. The three isomers of pentane are known as normal pentane, isopentane, and neopentane.

Another method of nomenclature is sometimes used for the simpler hydrocarbons in which the molecule is considered to be a derivative of methane. When the $-CH_3$ group occurs in a molecule, it is called a methyl group. Similarly, the $-C_2H_5$ group is known as the ethyl group. In general, a group of this type is known as an alkyl group and it is named by simply replacing the suffix -ane of the parent hydrocarbon with -yl. In order to name a hydrocarbon as a methane derivative the most highly substituted carbon atom (the carbon where the most branching of the chain occurs) is considered to be the original methane carbon and the alkyl groups attached to it are assumed to be replacing hydrogen atoms on the original methane molecule. As an example, for isopentane whose structure is shown before, the second carbon from the left is the most highly substituted so it is taken as the methane carbon. This methane can be considered as having two hydrogens replaced by methyl groups and the third hydrogen by an ethyl group. Consequently, another name for isopentane is dimethyl ethyl methane. Similarly, neopentane is called tetramethyl-methane. This method of nomenclature suffers from the disadvantage of not being generally applicable. It could not be used, for example, in the case of a highly branched hydrocarbon where no single carbon atom can be singled out as being the most highly substituted.

The Geneva system is a method of nomenclature which does not have this disadvantage since it is applicable to even the most complex hydrocarbons. To use this method it is first necessary to pick out the longest continuous chain of carbon atoms in the molecule and number them consecutively. The compound is considered to be a derivative of this longest continuous chain hydrocarbon. The various alkyl groups attached to this chain are named in the usual manner and their position is indicated by the number of the carbon atom to which they are joined. For example, for isopentane, there are four carbons in a

$$\begin{array}{l} CH_3CHCH_2CH_3 \\ \qquad\;\; | \\ \qquad\; CH_3 \end{array}$$

continuous chain so this isomer of pentane is considered to be a derivative of butane. A methyl group is attached to the second carbon in the longest chain. Consequently by the Geneva system this compound is named 2-methylbutane. Similarly, by the Geneva system, the name for neopentane is 2,2-dimethylpropane. An example of a more complex compound named by this method is the isomer of nonane, which

$$\begin{array}{l} CH_3CHCHCH_2CH_2CH_3 \\ \qquad\;\; | \quad\; | \\ \quad\; H_3C \;\; C_2H_5 \end{array}$$

is designated as 2-methyl-3-ethylhexane. It is conventional to number the longest chain so that the numbers appearing in the name are as small as possible. This is, however, a minor point and it is obvious that, no matter from which end the chain is numbered, the name will clearly indicate the structure of the molecule.

Chemical and Physical Properties of the Paraffins. The paraffins are very important constituents of crude oil. Pennsylvania petroleum seems to be largely composed of hydrocarbons of this series while other petroleums contain them to a lesser extent. Natural gas consists of the more volatile members of this series.

The paraffins are characterized by their chemical inertness. They will not react with concentrated sulfuric or nitric acid. However, when ignited in the presence of air or oxygen they burn giving off large amounts of heat and under proper conditions this combustion is explosive. However, the reaction with oxygen occurs only at elevated temperatures. The inertness of the paraffin hydrocarbons accounts for their presence in petroleum since their existence for geological periods of time would require a high degree of stability.

The first four members of this series are gases under standard conditions of temperature and pressure (60° F and 14.7 psia). Those from C_5H_{12} to $C_{17}H_{36}$ are liquids. From $C_{18}H_{38}$ upwards these hydrocarbons are colorless, waxlike solids. Paraffin is a mixture of the solid members of this series and it is from this fact that the series gets its name.

Unsaturated Hydrocarbons. The members of the paraffin series are characterized by the fact that the bonds between the carbons are single; one valence of each carbon is involved in bonding two carbons together. These hydrocarbons are said to be saturated since they contain all the hydrogen possible. Hydrocarbons exist, however, which have double or even triple bonds between carbons. Since these compounds will add more hydrogen under appropriate conditions they are said to be unsaturated.

The olefin series of hydrocarbons is characterized by the fact that one bond in the molecule is double, that is, two valences of each carbon joined by the double bond are involved. The simplest member of this series is ethylene whose structure is shown here.

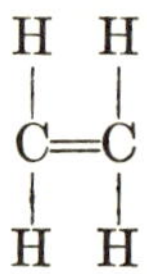

or $CH_2{=}CH_2$

The next member is propylene,

$$CH_2{=}CHCH_3$$

then butylene,

$$CH_2{=}CHCH_2CH_3$$

and so on. The general formula for this series is C_nH_{2n} which clearly indicates that these hydrocarbons contain less hydrogen than the corresponding members of the paraffin series. Nevertheless, the valence of carbon in these compounds, as in all hydrocarbons, is four.

According to the Geneva system these hydrocarbons are named in the same way as the paraffins except that the suffix -ene is used instead of -ane. However, the first three members of this series are more commonly referred to as ethylene, propylene, and butylene instead of ethene, propene, and butene. This is due to the survival of an older system of nomenclature. For the higher members of this series, the more modern suffix -ene is usually employed instead of the older -ylene.

Isomerism occurs in the olefins just as it does in the paraffins. Not only is it due to branching of the carbon chains but the position of the double bond in the molecule is also a source of isomerism. Thus, a molecule of butene, or butylene, with the double bond between the first and second carbons has properties quite different from that with the double bond between the second and third carbons. The Geneva system distinguishes these two isomers by adding a number at the end of the name. This number indicates the lowest numbered carbon atom involved in the double bond. Thus,

$$CH_2{=}CHCH_2CH_3$$

is known as butene-1, and

$$CH_3CH{=}CHCH_3$$

is butene-2.

An example of a more complicated olefin named by the Geneva system is

$$\begin{array}{l} CH_2{=}\underset{\substack{| \\ H_3C}}{C}\underset{\substack{| \\ C_2H_5}}{C}HCH_2CH_3 \end{array}$$

which is named 2-methyl-3-ethylpentene-1.

Another series of unsaturated hydrocarbons is known as the diolefins. They are characterized by the fact that there are two double bonds in the molecule. The general formula for this series is C_nH_{2n-2}. To name these compounds by the Geneva system the suffix -adiene is used and the positions of the two double bonds are indicated by two numbers placed after the name. For example,

$$CH_2{=}CHCH{=}CH_2$$

is known as butadiene-1,3, and

$$CH_2{=}C{=}CHCH_3$$

is called butadiene-1,2.

A third series of unsaturated hydrocarbons of considerable importance is the acetylene series. These compounds have a triple bond. The first member of this series has the structure shown here, and is

$$CH \equiv CH$$

commonly called acetylene. The Geneva system for naming these hydrocarbons is to use the suffix -ine. Thus by the Geneva system the name for acetylene is ethine. The general formula for this series is the same as that for diolefins, namely C_nH_{2n-2}.

Chemical and Physical Properties of the Unsaturated Hydrocarbons. The unsaturated hydrocarbons are, in contrast with the paraffins, very reactive. They react rapidly with chlorine to form oily liquids; hence the name olefin ("oil-forming"). Under the proper conditions they react readily with hydrogen which saturates the double bonds and forms the corresponding paraffin. Because of their high reactivity these unsaturated hydrocarbons are not found in crude oil to any great extent. However, they are formed in large amounts in petroleum-cracking processes and have considerable industrial importance.

The Naphthene Hydrocarbons. The naphthene hydrocarbons are also called cycloparaffins and, as this name implies, they are saturated hydrocarbons in which the carbon chains form closed rings. The general formula for this series is C_nH_{2n} and consequently they are isomeric with the olefins. They are named by placing the prefix cyclo- before the name of the corresponding paraffin hydrocarbon. The first member of this series is cyclopropane and has the structure shown. The

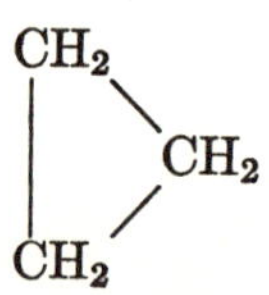

series continues with cyclobutane, cyclopentane, cyclohexane, and so on. If the carbon ring has side chains the compounds are named by

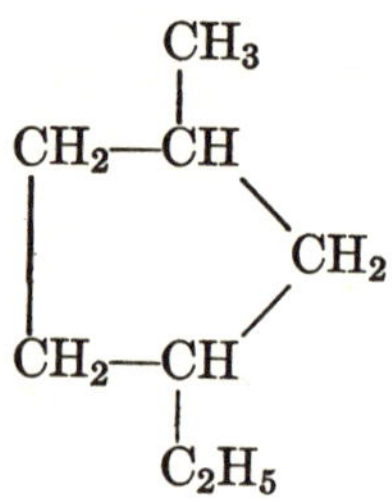

numbering the carbons in the ring and indicating the position of the substituting alkyl groups in the usual manner. Thus, the compound shown is named 1-methyl-3-ethylcyclopentane.

These compounds, being saturated, are very stable and are important constituents of crude oil. In general, the chemical properties of these hydrocarbons are very similar to those of the paraffins.

The Aromatic Hydrocarbons. These hydrocarbons are also cyclic and may be considered to be derivatives of benzene. Benzene has the formula C_6H_6, and the structure is as depicted. It consists of a six-

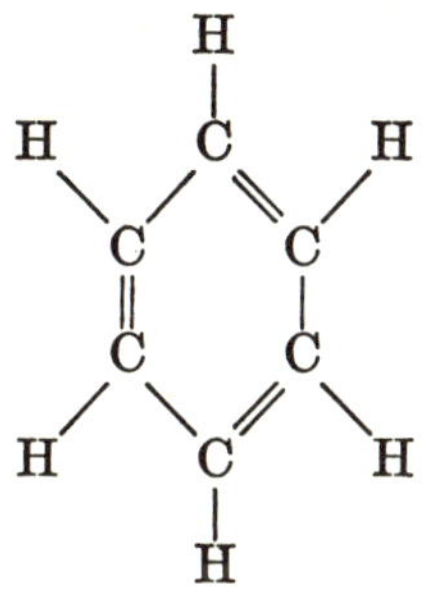

membered ring with alternate single and double bonds. This structure is so common in organic compounds that chemists use a hexagon as a special symbol to represent the benzene molecule. Some of the simpler members of this series consist of benzene with one or more alkyl groups as side chains. An example is methylbenzene whose structure is shown here. This particular hydrocarbon is also known as toluene,

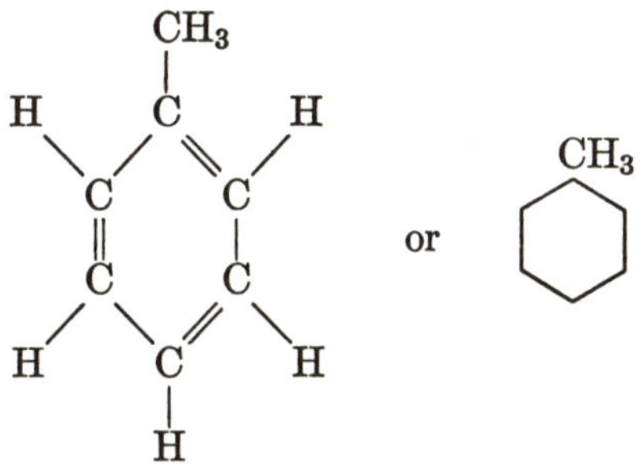

it being of sufficient importance to warrant a common name. If two or more alkyl groups are attached to the benzene ring as side chains, the carbons in the ring are numbered and the relative positions of the substituting groups are indicated in the usual manner. Thus,

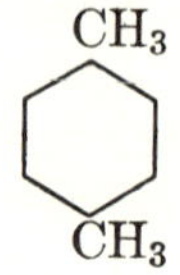

is known as 1,4-dimethylbenzene. The common name for this hydrocarbon is para xylene. The prefix para in this name indicates that the two methyl groups are attached to the carbon atoms occupying opposite positions in the benzene ring. There are two other isomers of dimethylbenzene which are designated as 1,2-dimethylbenzene and 1,3-dimethylbenzene. In the first of these compounds, which is also known as ortho xylene, the methyl groups occupy adjacent positions in the benzene ring while in the second, which is also known as meta xylene, the methyl groups are attached to carbon atoms which are separated by one carbon atom. Note that the general formula for the hydrocarbons of this series is C_nH_{2n-6}.

The fact that the benzene ring contains three double bonds suggests that the members of this series should be very reactive. However, this is not so, and although they are not as stable as the paraffins, they do not show the high reactivity that is so characteristic of the olefins. The reason for this apparent anomaly has troubled chemists for over a half century and it has not been until the relatively recent application of quantum mechanical concepts that this problem has been satisfactorily answered. However, a thorough discussion of the problem is beyond the scope of this book and the student is asked to accept the stability of benzene without a simple explanation for this fact. Compounds of this series do occur in crude oil. Indeed, petroleum is one of the important sources of these very important hydrocarbons.

The aromatic hydrocarbons are either liquids or solids under standard conditions of temperature and pressure. Benzene is a colorless liquid which boils at 176° F. Many of the members of this series are characterized by fragrant odors; hence the name "aromatic" is given to this series.

Petroleum Oil. Petroleum oil or crude oil is a complex mixture consisting largely of hydrocarbons belonging to the various series that have been described. In addition, crude oils usually contain small amounts of combined oxygen, nitrogen, and sulfur. No crude oil has ever been entirely separated into its individual components. Indeed, with present day techniques of separation and analysis this would be an almost impossible undertaking. However, a large number of hydrocarbons have been separated from crude oil and identified. A partial listing of some of the more important compounds that have been isolated from a crude oil taken from the South Ponca field in Oklahoma are listed in Table 1. All told, a total of 141 compounds,

Table 1

Paraffins	Naphthenes	Aromatics
All normal paraffins to $C_{10}H_{22}$	Cyclopentane	Benzene
Isobutane	Cyclohexane	Toluene
2-Methylbutane	Methylcyclopentane	Ethylbenzene
2,3-Dimethylbutane	1,1-Dimethylcyclopentane	Xylene
2-Methylpentane	Methylcyclohexane	1,2,4-Trimethylbenzene
3-Methylpentane	1,3-Dimethylcyclohexane	
2-Methylhexane	1,2,4-Trimethylcyclohexane	
3-Methylhexane		
2-Methylheptane		
2,6-Dimethylheptane		
2-Methyloctane		

accounting for about 44% of the crude oil volume, have been isolated and identified in this particular oil sample.

Crude oils obtained from various localities have widely different characteristics indicating that the constituent hydrocarbons have different properties. However, if an ultimate analysis is made for carbon, hydrogen, sulfur, nitrogen, and oxygen on a wide variety of crude oils the results will show only minor differences. A mobile, low specific gravity oil will give essentially the same analysis as a viscous, pitch-like, high specific gravity oil. Nearly all crude oils will give ultimate analyses within the following limits:

Element	Percentage by Weight
Carbon	84–87
Hydrogen	11–14
Sulfur	0.06–2.0
Nitrogen	0.1–2.0
Oxygen	0.1–2.0

This constancy may at first seem remarkable, but it is to be expected if one keeps in mind the fact that hydrocarbons are essentially long chains of CH_2 groups. Consider two hypothetical crude oils, one consisting of pure octane (C_8H_{18}) and the second of pure dodecane ($C_{12}H_{26}$). These two "crudes" would have entirely different physical and chemical properties. The first would be an excellent gasoline, whereas the second would have the properties of kerosene. However, the percentages of carbon and hydrogen in each would be very much the same as is shown by the following simple calculations. For octane (C_8H_{18}) (molecular weight = 114)

$$\text{Per cent H} = \frac{\text{Wt of hydrogen per molecular weight}}{\text{Molecular weight}} = \frac{18 \times 1}{114} \times 100$$

$$= 15.8 \text{ per cent}$$

$$\text{Per cent C} = \frac{\text{Wt of carbon per molecular weight}}{\text{Molecular weight}} = \frac{8 \times 12}{114} \times 100$$

$$= 84.2 \text{ per cent}$$

For dodecane ($C_{12}H_{26}$) (molecular weight = 170)

$$\text{Per cent H} = \frac{26 \times 1}{170} \times 100 = 15.3 \text{ per cent}$$

$$\text{Per cent C} = \frac{12 \times 12}{170} \times 100 = 84.7 \text{ per cent}$$

Natural Gas. Natural gas can occur by itself or in conjunction with liquid petroleum oils. It consists mainly of the more volatile members of the paraffin series containing from one to four carbon atoms per molecule; however, it is understood that small amounts of higher molecular weight hydrocarbons can be present. In addition, natural gases may contain varying amounts of carbon dioxide, nitrogen, hydrogen sulfide, helium, and water vapor. Most natural gases consist predominantly of methane, the percentage of which may be as high as 98 per cent. Natural gas can be classified as sweet or sour and as wet or dry. A sour gas is one which contains appreciable amounts of hydrogen sulfide and consequently can be quite corrosive. The designation wet gas has nothing to do with the presence of water vapor but signifies that the gas will yield appreciable quantities of liquid hydrocarbons with proper treatment. Water vapor is, however, often present in natural gas and sometimes causes stoppages in high-pressure gas lines during cold weather. This is due to the fact that hydrocarbons form solid hydrates with water at high pressure and low temperature.

Tars and Asphalts. These solid and semisolid substances are also known as bitumens, waxes, resins, and pitch. They are very complex substances and relatively little is known regarding their chemical composition. There is little doubt that these materials were formed in nature from petroleum oils by evaporation of the more volatile constituents, and oxidation and polymerization of the residue. Indeed, products closely resembling the natural materials can be prepared by heating and air blowing crude oil.

Products from Petroleum. The distillation of crude oil results in various fractions which boil at different temperatures. The lowest boiling fraction which is collected from room temperature up to about 160° F consists largely of hydrocarbons in the range from C_5 to C_6. It is known as petroleum ether or ligroin and is used as a solvent. The fraction which boils from about 160° F to 400° F is known as gasoline. It consists of hydrocarbons containing from seven to eleven carbons. Kerosene is the next fraction which boils in the range 400° F to 575° F. The lubricating oils and fuel oils consist of hydrocarbons which boil above 575° F.

If the residue which remains after distillation is a waxlike solid consisting largely of paraffin hydrocarbons the crude is designated as paraffin base. If the residue is a black pitchlike solid the crude is called asphalt base. Pennsylvania crude has a paraffin base whereas California oils are for the most part asphalt base. Often a clear-cut distinction cannot be made and the crude is described as being mixed base oil. Most Mid-Continent crudes are of this type.

Today gasoline is one of the most important products of petroleum and consequently deserves a more thorough description than that given above. In the early days of petroleum production, kerosene was the product in demand because of its desirable characteristics as an illuminating oil. With the advent of the internal combustion engine, gasoline became increasingly important. The yield of gasoline from a crude can be increased by subjecting the less volatile constituents to a thermal decomposition so that they are broken down into smaller, more volatile products. This process is known as "cracking." The reactions which occur are complex and exceedingly varied but the following may be taken as typical:

$$C_{14}H_{30} \xrightarrow{750\text{–}1300^\circ F} C_7H_{16} + C_7H_{14}$$

This reaction indicates the decomposition of tetradecane, a constituent of kerosene into heptane and heptene. The yield of gasoline can be further augmented by extracting the less volatile constituents of natural gas. As previously pointed out, natural gas does contain some of the constituents of gasoline in small amounts. When these are recovered the product is a highly volatile liquid known as natural gasoline or casing-head gasoline, which can be mixed with regular gasoline to form an acceptable product. The recovery is usually accomplished by compression and cooling of the natural gas so that liquid condenses out or by absorption on charcoal or in oil from which the natural gasoline can subsequently be recovered by heating.

The fractions obtained by distilling crude oil are the raw materials for the manufacture of a wide variety of other useful products. Some of the important commercial products obtained from petroleum are listed in Figure 2.

Products from Petroleum

	Fractions Obtained by Distillation	Final Products Obtained by Redistillation and Further Treatment
Crude oil	Hydrocarbon gases	Natural gas, bottled fuel gas.
	Petroleum ether	Solvents, paint thinners, cleaners.
	Gasoline	Motor fuels, solvents, toluene for explosives, etc.
	Kerosene	Illuminating oil, diesel fuel, gas adsorption oils.
	Fuel oil	Heating fuels, diesel fuel, cracking stock, naphthenic acid which may be converted to lubricating oil additives, paint driers, fungicides, etc.
	Lubricating oil stock	Lubricants of all kinds, medicinal oil, transformer oil, wax, hydraulic oil, detergents and wetting agents, rust preventatives, waterproofing compounds.
	Residue	Wood preservatives, roofing compounds and shingle saturants, road oils, paving asphalt, insulating asphalt, coke.

FIG. 2. Diagram showing some of the important products obtained from petroleum.

REFERENCES

1. Any good textbook on organic chemistry. The following are suggested:
   Conant, James B., and A. H. Blatt, *The Chemistry of Organic Compounds,* The Macmillan Co., New York (1952).
   Fieser, Louis F., and Mary Fieser, *Organic Chemistry,* D. C. Heath & Co., Boston (1950).
   Gilman, Henry, *Organic Chemistry,* John Wiley & Sons, New York (1955).
2. Gruse, Wm. A., and Donald B. Stevens, *The Chemical Technology of Petroleum,* McGraw-Hill Book Co., New York (1942).
3. Nelson, W. L., *Petroleum Refinery Engineering,* McGraw-Hill Book Co., New York (1949).

PROBLEMS

1. Draw structural formulas for the isomers of heptane. Name the isomers by the Geneva system.
2. What are two other names for neopentane?

3. Draw structural formulas for the following compounds: (*a*) 2-methyl-4-propyloctane, (*b*) 2,4-dimethylhexene-3, (*c*) 2-methylbutadiene-1,3, (*d*) dimethylpropylmethane, (*e*) 1-methyl-2-ethylcyclohexane, (*f*) 1,2-diethylbenzene.

4. Define: (*a*) isomer, (*b*) diolefin, (*c*) cyclic hydrocarbon, (*d*) unsaturated hydrocarbon, (*e*) wet gas, (*f*) paraffin base crude, (*g*) asphalt.

5. What are the general formulas for following hydrocarbon series? (*a*) paraffin series, (*b*) olefin series, (*c*) diolefin series, (*d*) acetylene series, (*e*) naphthene series, (*f*) cycloparaffin series, (*g*) aromatic series.

CHAPTER 2

BEHAVIOR OF GASES

A gas may be defined as a homogeneous fluid, generally of low density and viscosity, that has no definite volume but fills completely any vessel in which it is placed. The laws which express the behavior of gases under various conditions of temperature and pressure are of utmost importance in petroleum technology. These laws are relatively simple for a hypothetical fluid known as a perfect gas and actual gases follow these laws to varying degrees of accuracy. The subject matter of this chapter will concern itself first with a discussion of the perfect gas laws and then with the behavior of actual gases which may deviate markedly from these laws under certain conditions of temperature and pressure.

THE PERFECT GAS LAWS

Boyle's Law is concerned with the effect of pressure on the volume of a gas. It may be stated: For a given weight of gas, at a given temperature, the volume varies inversely as the pressure. Mathematically Boyle's Law may be expressed

$$V \propto 1/P \qquad \text{or} \qquad PV = \text{Constant} \tag{1}$$

Consequently, if the volume of a given weight of gas is plotted as a function of pressure at constant temperature, the resulting curve will be a hyperbola.

Charles' Law (also known as Gay-Lussac's Law) deals with the effect of temperature on the volume of a gas and may be stated: For a given weight of gas, at a given pressure, the volume varies directly as the absolute temperature, or

$$V \propto T \qquad \text{or} \qquad V/T = \text{Constant} \tag{2}$$

The absolute temperature is expressed in degrees Rankine which is equal to the Fahrenheit temperature plus 460, or in degrees Kelvin which is equal to the Centigrade temperature plus 273. Thus,

$$\begin{aligned} {}^\circ\,\text{Rankine} &= ({}^\circ\,\text{R}) = {}^\circ\,\text{F} + 460 \\ {}^\circ\,\text{Kelvin} &= ({}^\circ\,\text{K}) = {}^\circ\,\text{C} + 273 \end{aligned} \tag{3}$$

According to Charles' Law, if the volume of a given quantity of gas at constant pressure is plotted as a function of the absolute temperature, a straight line will result. This line will pass through the origin indicating that at absolute zero the volume is zero. It is obvious, however, that long before absolute zero is reached, any actual gas will have liquefied and this simple law will no longer be applicable.

Charles' Law and Boyle's Law can be combined to describe the behavior of a gas when both temperature and pressure are altered. Assume a given weight of gas whose volume is V_1 at P_1 and T_1. Imagine the following process whereby the gas reaches the volume V_2 at P_2 and T_2.

$$\begin{array}{ccc} (V_1P_1T_1) & \xrightarrow[T_1=\text{Constant}]{\text{Step 1}} & (VP_2T_1) \\ & & \text{Step 2}\downarrow P_2=\text{Constant} \\ & & (V_2P_2T_2) \end{array}$$

In the first step the pressure is changed from P_1 to P_2, keeping the temperature constant. In step 2 the pressure is kept constant at P_2 and the temperature is altered from T_1 to T_2. Since the quantity of gas and the temperature are constant in the first step, Boyle's Law applies and one may write

$$P_1V_1 = P_2V \qquad \text{or} \qquad V = P_1V_1/P_2$$

where V represents the volume at pressure P_2 and temperature T_1. Since Charles' Law applies to the second step one has

$$V/T_1 = V_2/T_2 \qquad \text{or} \qquad V = V_2T_1/T_2$$

Eliminating V between these two equations gives

$$P_1V_1/P_2 = V_2T_1/T_2 \qquad \text{or} \qquad P_1V_1/T_1 = P_2V_2/T_2 \tag{4}$$

In other words, for a given weight of gas, PV/T = constant.

The third perfect gas law to be considered is known as Avogadro's Law. It can be stated: Under the same conditions of temperature and pressure, equal volumes of all perfect gases contain the same number of molecules. This is equivalent to the statement that one molecular weight of any perfect gas occupies the same volume at a given temperature and pressure. Thus, one molecular weight in grams of any gas at 0° C and one atmosphere pressure occupies a volume of 22.4 liters. Similarly, one molecular weight in pounds of any gas at 60° F and 14.7 psia occupies a volume of 379 cu ft. It is customary to define the above conditions of temperature and pressure as *standard*. Consequently, according to Avogadro's Law one lb mole of any perfect gas occupies a volume of 379 standard cu ft and one g mole occupies a volume of 22.4 standard liters.

If one combines Boyle's Law and Charles' Law with Avogadro's Law and considers one molecular weight of gas, equation 4 becomes

$$PV/T = R$$

where R is a constant that has the same value for *all* gases. For n moles of gas this equation becomes

$$PV = nRT \tag{5}$$

and since n is the weight of gas divided by the molecular weight, this equation can be written

$$PV = \frac{\text{Wt}}{\text{MW}} RT \tag{6}$$

This expression is known as the general gas law and describes the behavior of a perfect gas. No gas is perfect, and all actual gases deviate more or less from this simple law. However, the behavior of perfect gases will be discussed more fully before considering any deviations that may occur.

The value of the constant, R, obviously depends on the system of units employed to express temperature, pressure, and volume. Suppose the pressure is expressed in atmospheres, volume in liters, temperature in degrees Kelvin, and moles in gram-moles. In this system of units Avogadro's Law states that one gram-mole of any gas occupies 22.4 liters at 273° K and one atm pressure. Therefore

$$R = \frac{PV}{nT} = \frac{1 \times 22.4}{1 \times 273} = 0.0821 \text{ liter atm per degree per gram-mole}$$

In this manner a value of R can be calculated for any system of units.

Table 2 gives the value of R in the systems of units most commonly employed in engineering calculations.

Table 2. Value of the Gas Constant

P	V	T	n	R
atm	liters	° K	$\frac{\text{grams}}{\text{MW}}$	0.0821
atm	cc	° K	$\frac{\text{grams}}{\text{MW}}$	82.1
psia	cu ft	° R	$\frac{\text{lb}}{\text{MW}}$	10.72
psfa	cu ft	° R	$\frac{\text{lb}}{\text{MW}}$	1544

Example. Four pounds of methane are placed in a tank at 60° F. If the pressure on the tank is 100 psia what is the volume of the tank? Substitution in the general gas law (equation 5) yields

$$V = \frac{nRT}{P} = \frac{\frac{4}{16} \times 10.72 \times (460 + 60)}{100} = 13.95 \text{ cu ft}$$

Since the value of 10.72 was used for R it is essential that n be in lb moles, T in ° R, P in psia. The units of V will consequently be cu ft.

The Density of a Perfect Gas. Since density is defined as the weight per unit volume, the general gas law can be used to calculate densities of gases at various temperatures and pressures. Solving the general gas law for Wt/V

$$\frac{\text{Wt}}{V} = \frac{\text{MW} \times P}{RT} = \text{Density} = D_g \tag{7}$$

If R is taken to be 10.72, and P and T are expressed in psia and ° R respectively, the units of density will be lb/cu ft.

Gas Mixtures. Usually the engineer is interested in the behavior of gas mixtures and seldom deals with gases that consist of only one component. The compositions of gas mixtures are commonly expressed as weight %, volume %, or mole %. The weight % of a particular component is defined as the weight of that component divided by the total weight multiplied by 100 so the result will be on a percentage basis. Thus for the ith component of weight Wt_i

$$(\text{Weight }\%)_i = \frac{\text{Wt}_i}{\Sigma \text{Wt}} \times 100 \tag{8}$$

where ΣWt_i represents the total weight of the system. Similarly the volume % of the ith component is defined as

$$(\text{Volume }\%)_i = \frac{V_i}{\Sigma V_i} \times 100 \tag{9}$$

where V_i represents the volume of the ith component, and ΣV_i is the total volume. $(\text{Mole }\%)_i$ is defined as

$$(\text{Mole }\%)_i = \frac{n_i}{\Sigma n_i} \times 100 \tag{10}$$

Instead of mole %, the term "mole fraction" is sometimes used. The mole fraction of the ith component is

$$y_i = \frac{n_i}{\Sigma n_i}$$

The concepts of weight % and volume % are self-explanatory. However, the concept of mole % or mole fraction should be described more fully so that its meaning is clearly understood. Basically, the mole fraction represents the fraction of molecules in the system that are of a given kind. This follows at once from the fact that one mole of any gas contains the same number of molecules. For example, suppose a system contains one mole of CH_4 and two moles of C_2H_6. In this system the mole fraction of CH_4 is ⅓ and that of C_2H_6 is ⅔. It is also true that ⅓ of the molecules are CH_4 molecules and ⅔ are C_2H_6 molecules.

Relationship between Mole % and Volume %. If each gas in a mixture obeys Avogadro's Law, the volume of the ith component would be proportional to the number of moles of the ith component. That is,

$$V_i \propto n_i \qquad \text{or} \qquad V_i = kn_i$$

where k is the proportionality constant. Substituting the above values for V_i into the definition of $(\text{Volume }\%)_i$ it is seen that

$$(\text{Volume }\%)_i = \frac{V_i}{\Sigma V_i} \times 100 = \frac{kn_i}{\Sigma kn_i} \times 100 = \frac{n_i}{\Sigma n_i} \times 100 = (\text{Mole }\%)_i$$

It follows, therefore, that, for gases which obey Avogadro's Law, volume % and mole % are equivalent.

Relationship between Weight % and Mole % (or Volume %). The process of conversion from weight % to mole % can best be explained by a definite example. Assume a gas mixture whose components are given in column 1 in the table below. The weight % of each component is listed in column 2. Assume 100 lb of the gas mixture as a basis. In column 3 the actual weight of each component is listed. The number of moles (n_i) of each component is given in column 4 and is obtained by dividing the weight of each component by its molecular weight. The total number of moles in the system, Σn_i, is the sum of the figures in column 4. The mole % of each component is listed in column 5 and is obtained by dividing n_i by Σn_i and multiplying by 100 as required by definition. It is obvious that the results

(1) Components	(2) Wt %	(3) Wt_i per 100 lb	(4) Moles (n_i) per 100 lb	(5) Moles % $= \frac{n_i}{\Sigma n_i} \times 100$
CH_4	60.0	60	$\frac{60}{16} = 3.750$	77.86%
C_2H_6	20.0	20	$\frac{20}{30} = 0.667$	13.85%
C_3H_8	10.0	10	$\frac{10}{44} = 0.227$	4.71%
C_4H_{10}	10.0	10	$\frac{10}{58} = 0.172$	3.57%
			$\Sigma n_i = 4.816$	

would be the same no matter what weight of gas had been taken as the basis. One hundred pounds was chosen since this simplifies the calculation.

The conversion from mole % (or volume % since they are identical) to weight % will be explained by an example which is the reverse of the preceding conversion. The components are listed in column 1 below. The mole % of each component is listed in column 2. One mole of the mixture is chosen as a basis. The weight of each component is tabulated in column 3 and is obtained by multiplying the number of moles of each component by its molecular weight. The sum of the figures in column 3 will represent the total weight of the system (i.e., weight of one mole). The weight % of each component is given in column 4. Here again, any quantity of gas could have been chosen

(1) Components	(2) Mole %	(3) Weight (Wt_i) (basis = one mole)	(4) Weight % $= \frac{Wt_i}{\Sigma Wt_i} \times 100$
CH_4	77.86	$0.7786 \times 16 = 12.46$	60.0%
C_2H_6	13.85	$0.1385 \times 30 = 4.16$	20.0%
C_3H_8	4.71	$0.0471 \times 44 = 2.07$	10.0%
C_4H_{10}	3.57	$0.0357 \times 58 = 2.07$	10.0%
		$\Sigma Wt_i = 20.76$	

as a starting basis and the final results would have been the same.

The Concept of Apparent Molecular Weight. It is not proper to speak of the molecular weight of a mixture. However, a gas mixture *behaves* as though it had a definite molecular weight. It has previously been stated that 1 lb molecular weight of any perfect gas has a volume of 379 cu ft at 60° F and 14.7 psia. Conversely, the weight of 379 cu ft of a gas mixture at 60° F and 14.7 psia is called the apparent molecular weight. The gas mixture behaves as though it were a pure gas whose molecular weight was equal to the apparent molecular weight. If y_i represents the mole fraction of the ith component in a gas mixture the apparent molecular weight can also be defined as

$$\text{AMW} = \Sigma(y_i \times \text{MW}_i) \tag{11}$$

since this quantity represents the weight of one mole.

EXAMPLE. Calculate the apparent molecular weight of a gas mixture consisting of three moles of methane, one mole of ethane and one mole of propane. The mole fractions of methane, ethane and propane in this mixture are 0.60, 0.20, and 0.20, respectively. Consequently

$$\text{AMW} = \Sigma(y_i \times \text{MW}_i) = 0.60 \times 16 + 0.20 \times 30 + 0.20 \times 44 = 24.4$$

The concept of apparent molecular weight is a very useful one since it permits the general gas law to be applied to gas mixtures, provided the molecular weight in the gas law is replaced by the apparent molecular weight.

Dry air is a gas mixture consisting essentially of nitrogen, oxygen, and small amounts of other gases. Its composition is given in the table below

Table 3. Composition of Dry Air

Component	Mole Fraction
Nitrogen (N_2)	0.78
Oxygen (O_2)	0.21
Argon (A)	0.01

Application of equation 11 leads to a value of 28.96 for the apparent molecular weight of air. However, for most engineering calculations a value of 29.0 is considered to be sufficiently accurate.

Specific Gravity of a Gas. Specific gravity is defined as the ratio of the density of a substance to the density of some standard substance. In the case of liquids and solids, water is usually taken as the standard reference material and its density taken at 20° C and one atmosphere. In engineering units the density of water is usually taken at 60° F and one atmosphere. To avoid ambiguity the temperature and pressure of the reference substance should always be specified.

For gases the standard reference material is dry air and its density is taken at the same temperature and pressure for which the density of the gas is given. The distinction between specific gravity and density must be kept clearly in mind. In the metric system, the density of water is essentially equal to one. Consequently, specific gravities and densities have substantially the same numerical values in this system of units. However, this is *not generally* true.

The specific gravity of a gas is defined as

$$\text{S.G.} = \frac{D_g}{D_a} \tag{12}$$

where D_g and D_a are the densities of the gas and air respectively. Keeping in mind that for a perfect gas the density is given by

$$D_g = \frac{\text{MW} \times P}{RT}$$

and, assuming that the gas and air are both perfect gases, one may write

$$\text{S.G.} = \frac{D_g}{D_a} = \frac{\text{MW} \times P/RT}{\text{AMW}_a \times P/RT} = \frac{\text{MW}}{29}$$

since the two densities are measured at the same temperature and pressure. If the gas is a mixture, this equation obviously becomes

$$\text{S.G.} = \frac{\text{AMW}}{29} \tag{13}$$

Experimental Determination of Gas Gravity. For the laboratory determination of specific gravity of a gas one has recourse to several experimental methods. From equation 12 it is apparent that the specific gravity of a gas is equal to the ratio of the mass of a given volume of gas to the mass of an equal volume of dry air measured at the same temperature and pressure. This concept is the basis for the direct weighing method of measuring gas gravity in the laboratory. The weight of an evacuated glass bulb fitted with inlet and outlet stopcocks is determined on an analytical balance. Subsequently, the weight of bulb is determined again, first when filled with the unknown gas and then with dry air. The weight of the unknown gas divided by the weight of the dry air will be equal to the gas gravity, provided the weights of the gases were determined at the same temperature and pressure.

Instead of weighing the gases directly, the buoyant force of a gas acting on a sealed glass bulb suspended in it can be measured. Since the buoyant force is directly proportional to the gas density and density in turn is directly proportional to the pressure, it follows that the buoyant force of a gas varies directly as the pressure. Consequently, if the pressures of an unknown gas and dry air are adjusted so that both gases exert the same buoyant force on a given bulb their densities will vary inversely as their pressures. From equation 12 it follows that

$$\text{S.G.} = \frac{P_{\text{air}}}{P_{\text{gas}}} \tag{14}$$

where P_{air} is the absolute pressure of air required to exert a given buoyant force on the bulb and P_{gas} is the absolute pressure of the unknown gas required to exert the same buoyant force on the same bulb.

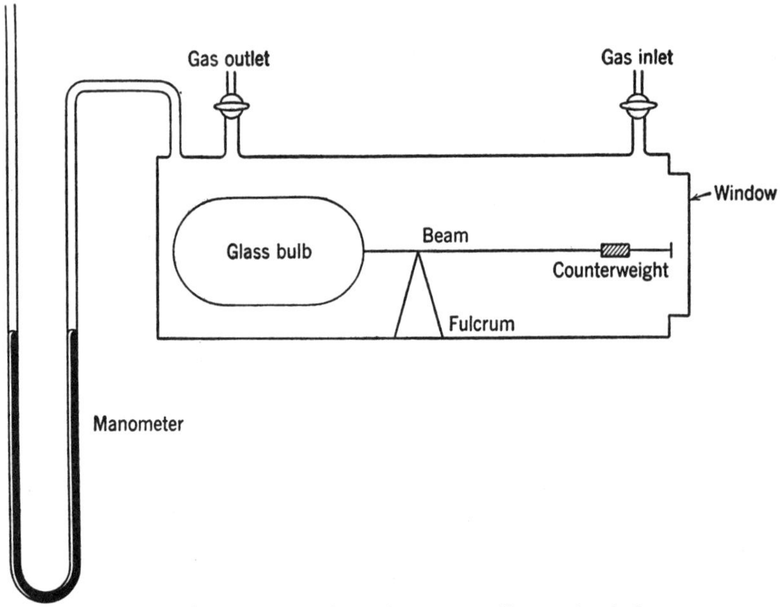

FIG. 3. Diagrammatic view of a gas-specific gravity balance.

Instruments designed to measure the buoyant force of a gas usually consist of a beam mounted on a fulcrum (Figure 3). On one end of the beam is attached a sealed glass bulb whose volume is large compared to the counterweight on the other end. This beam is contained in a gas-tight chamber with a windowed end which permits the beam to be observed. By altering the pressure of the gas in the chamber the beam can be balanced. A manometer attached to the chamber allows the pressure within to be measured.

In addition to these weighing methods of determining gas gravity, a method based on the effusion of gases through a small orifice can be employed. This method is based on Graham's Law which may be stated: The rate of effusion of a gas through an orifice is inversely proportional to the square root of the gas density. Consequently, the

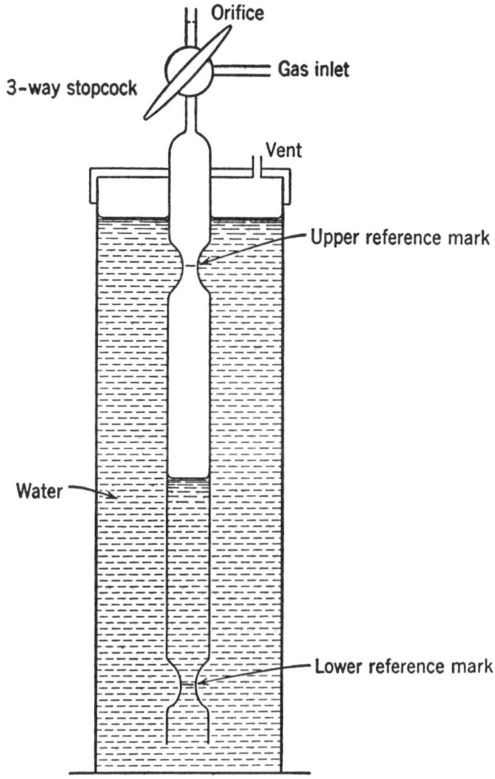

FIG. 4. Diagrammatic view of apparatus for measuring gas gravity by the effusion method.

ratio of the times required for the flow of equal volumes of two gases under the same pressure through a given orifice is equal to the ratio of the square root of their densities. If one of the gases is air the following equation may be written

$$\frac{t_{\text{gas}}}{t_{\text{air}}} = \sqrt{\frac{D_g}{D_a}} = \sqrt{\text{S.G.}} \tag{15}$$

where t_{air} and t_{gas} are the effusion times for equal volumes of air and gas through a given orifice under the same pressure.

An apparatus to measure gas gravity by this method is shown in Figure 4. A glass tube, open at the bottom and having a small orifice at the top, is held in a fixed position in a second, larger glass vessel. This vessel is filled nearly to the top with water. Gas or air is forced through a 3-way stopcock at the top and allowed to bubble out through the bottom of the inner glass tube until the water is saturated with the gas. The source of gas is removed and the gas contained in the inner tube is allowed to be displaced upward through the orifice by the pressure head exerted by the water. As the water rises in the inner tube the time required for the water to rise between two reference marks on the inner tube is recorded.

This simple method of measuring gas gravity is often used in the field. However, the results obtained by this method are only approximate and, if more accurate results are required, one of the weighing techniques previously described can be employed. The student should recognize that the effusion method employs gases saturated with water vapor and a correction must be made to obtain the water-free gas gravity.

Dalton's Law of Partial Pressures. This law may be stated: In a mixture of gases, each gas exerts a partial pressure equal to the pressure it would exert if it alone were present in the volume occupied by the mixture. In other words, the total pressure of a mixture of gases is equal to the sum of the partial pressures of its components. For example, consider two vessels of equal volumes containing methane and ethane respectively, each at a pressure of one atmosphere. If the ethane were forced into the methane tank at constant temperature, the pressure of the combined gases would be two atmospheres. According to Dalton's Law the partial pressures of the methane and ethane in the mixture would each be one atmosphere.

If perfect gas behavior is assumed partial pressures can be calculated in the following manner. Suppose the mixture contains n_1 moles of component 1, n_2 moles of component 2, n_3 moles of component 3, and so on. The total number of moles in the mixture is Σn_i. According to the general gas law, the pressure of the mixture in volume V and at temperature T is equal to

$$P = \Sigma n_i \frac{RT}{V}$$

Similarly, the partial pressure of component i in the same volume and at the same temperature is

$$P_i = n_i \frac{RT}{V}$$

Obviously

$$\frac{P_i}{P} = \frac{n_i}{\Sigma n_i} = y_i \qquad \text{or} \qquad P_i = y_i \times P \tag{16}$$

Consequently, the partial pressure of a gas in a mixture is given by the product of its mole fraction and the total pressure.

NON-PERFECT GASES

The Behavior of Non-Perfect Gases. *Van der Waals' Equation.* Actual gases approach perfect gas behavior at high temperatures and low pressures. In a perfect gas the molecules themselves are assumed to be of negligible volume (compared to the volume of the gas) and to exert no attractive forces on one another. At high pressures and low temperatures this is not so since, under these conditions, the volume of the molecules themselves is no longer negligible and the molecules, being more closely packed, exert appreciable attractive forces on one another.

Van der Waals' equation is an attempt to modify the general gas law so that it will be applicable to non-perfect gases. The equation for *one mole* of a single, pure gas is written

$$\left(P + \frac{a}{V^2}\right)(V - b) = RT \tag{17}$$

where a and b are constants whose values are different for each gas. The quantity a/V^2 accounts for the attractive forces between the molecules. It is added to the pressure because the actual pressure would need to be larger to produce the same volume than if no attraction existed. The constant b represents the volume of the molecules themselves, and it is subtracted from V since the actual volume of space available to the gas is less than the overall volume of the gas. When V is large (at low pressure and high temperatures), it is obvious that Van der Waals' equation reduces to the general gas law

$$PV = RT$$

Table 4 gives the values of a and b for several common gases. When using these constants it is necessary to express P in atmospheres, V in liters, T in ° K, and $R = 0.08205$.

If n moles of gas are involved it is apparent that equation 17 becomes

$$\left(P + \frac{n^2a}{V^2}\right)(V - nb) = nRT \tag{18}$$

Table 4. Van der Waals' Constants

Gas	*a* (atm liters^2)	*b* liters
CH_4	2.253	0.04278
C_2H_6	5.489	0.06380
C_2H_4	4.471	0.05714
C_2H_2	4.390	0.05136
CO_2	3.592	0.04267

If the constants a and b are not known it is possible to estimate their values from critical data. It can be shown (see for example: Daniels, *Outlines of Physical Chemistry*) that $a = 3P_cV_c^2$ and $b = V_c/3$ where P_c and V_c are the critical pressure and critical volume, respectively. The critical pressure is the pressure required to liquefy a pure gas at the critical temperature and the critical volume is the volume of one mole at the critical pressure and temperature. The meaning of these critical quantities will be described more fully in a later chapter.

Van der Waals' equation is not well suited to the calculations usually made by the engineer. Most often he has available pressure and temperature data and wishes to calculate a volume. To solve Van der Waals' equation for V involves the solution of a cubic equation, which is inconvenient. Furthermore, as previously pointed out, the engineer deals primarily with gas mixtures to which Van der Waals' equation is not readily applicable.

Consequently, a second method of treating imperfect gases will be described which applies equally well to single gases and to gas mixtures.

The Compressibility Factor. For an imperfect gas one can write the general gas law in the form

$$PV = ZnRT \tag{19}$$

where Z is known as the compressibility factor. It is an empirical factor, determined experimentally, which makes the above equation true at a particular temperature and pressure. For a perfect gas, Z is equal to one. For an imperfect gas, Z is greater or less than one, depending on the pressure and temperature. At a given temperature, the Z factor plotted as a function of pressure usually takes the form shown in Figure 5. There will be a similar curve for each value of temperature. Charts of this kind are available giving Z as a function of temperature and pressure for various hydrocarbons. Z as a function of temperature and pressure for methane, ethane, and propane are given in Figures 6, 7, and 8.

If a chart is not available for the particular gas in question a value of Z can still be estimated with reasonable accuracy. Use is made of the Law of Corresponding States which, for this purpose, may be

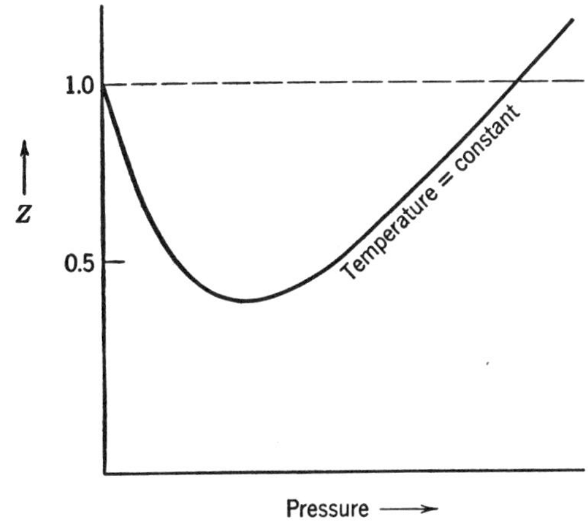

FIG. 5. Typical plot of the Z factor as a function of pressure at constant temperature.

stated: At the same reduced temperature and pressure, all hydrocarbons have the same value of Z. The reduced pressure and temperature are defined as

$$P_R = \frac{P}{P_c} \quad \text{and} \quad T_R = \frac{T}{T_c}$$

where T_c is the critical temperature and P_c is the critical pressure.

Table 5. Physical Properties of the Constituents of Natural Gas

Compound	Formula	MW	Critical Temperature °F	Critical Temperature °R	Critical Pressure, psia
Methane	CH_4	16.04	−116	344	673
Ethane	C_2H_6	30.07	89	549	712
Propane	C_3H_8	44.09	206	666	617
n-Butane	C_4H_{10}	58.12	306	766	551
Isobutane	C_4H_{10}	58.12	272	732	544
n-Pentane	C_5H_{12}	72.15	386	846	485
Isopentane	C_5H_{12}	72.15	370	830	483
n-Hexane	C_6H_{14}	86.17	454	914	435
n-Heptane	C_7H_{16}	100.20	512	972	397
n-Octane	C_8H_{18}	114.22	564	1024	362
Carbon dioxide	CO_2	44.01	88	548	1073
Nitrogen	N_2	28.02	−233	227	492
Hydrogen sulfide	H_2S	34.08	213	673	1306

Values for the critical temperature and pressure of several hydrocarbons are given in Table 5. To calculate T_R and P_R all temperatures and pressures must be in absolute units.

Figure 9 gives the Z factor as a function of reduced temperature and pressure. It has been prepared using experimental data for methane and expressing the temperature and pressure as reduced values. It is assumed that this chart applies to all pure hydrocarbon gases. This assumption is valid provided the Law of Corresponding States is applicable. Since hydrocarbon gases are closely related chemically, one would expect this law to hold reasonably well. However, it should be emphasized that reduced temperature and pressure Z factor charts prepared using experimental data for various hydrocarbon gases are not in exact agreement.

EXAMPLE. Calculate the volume of 10 lb of ethane at 145° F and 1068 psia. The critical pressure for ethane is 712 psia and its critical temperature is 549° R (Table 5). P_R and T_R are

$$P_R = \frac{P}{P_c} = \frac{1068}{712} = 1.50 \qquad T_R = \frac{T}{T_c} = \frac{460 + 145}{549} = 1.10$$

From Figure 9, Z is found to be 0.460 at this reduced temperature and pressure. Therefore

$$V = \frac{ZnRT}{P} = \frac{0.460 \times \frac{10}{30} \times 10.72 \times (460 + 145)}{1068} = 0.930 \text{ cu ft}$$

It is interesting to note that the Z factor actually determined experimentally for ethane at this temperature and pressure (see Figure 7) is 0.465. Assuming perfect gas behavior the calculated volume would have been 2.024 cu ft.

This method of correcting for non-ideal gas behavior can be extended to gas mixtures if the concepts of pseudo-critical pressure and pseudo-critical temperature are introduced. These quantities are defined

$$\text{pseudo-critical pressure} = \underline{P_c} = \Sigma(y_i \times P_{ci}) \tag{20}$$

$$\text{pseudo-critical temperature} = \underline{T_c} = \Sigma(y_i \times T_{ci}) \tag{21}$$

These pseudo-criticals (or molecular average criticals) are used in exactly the same way as the actual criticals are used for single gases. However, since the Law of Corresponding States is not an exact law, it is customary, when dealing with hydrocarbon gas mixtures, to use a Z factor chart, such as Figure 10, prepared from experimental data obtained from gas mixtures.

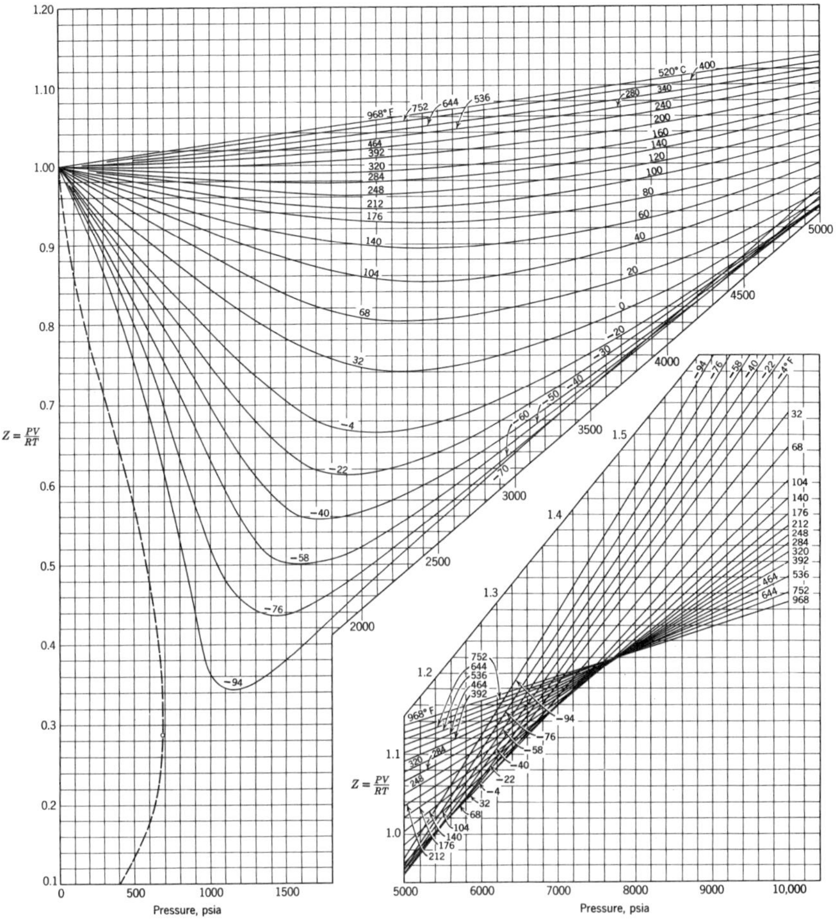

Fig. 6. Compressibility factors for methane. (Brown, Katz, Oberfell, and Alden, *Natural Gasoline and Volatile Hydrocarbons,* 1948, p. 24.)

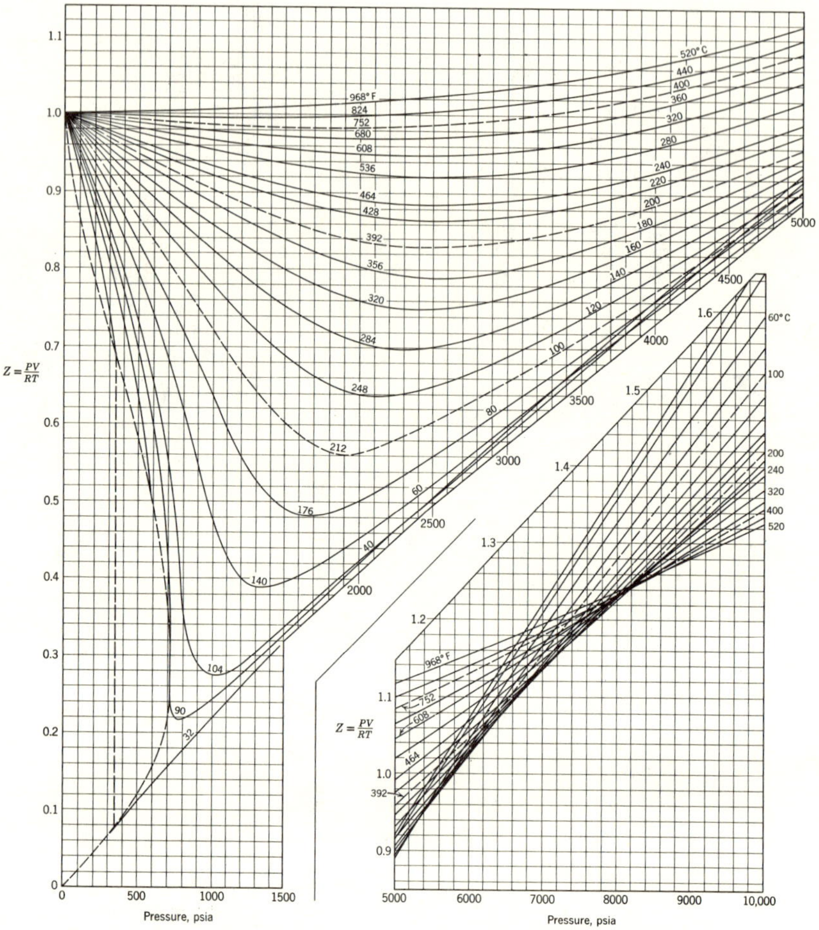

Fig. 7. Compressibility factors for ethane. (Brown, Katz, Oberfell, and Alden, *Natural Gasoline and Volatile Hydrocarbons,* 1948, p. 26.)

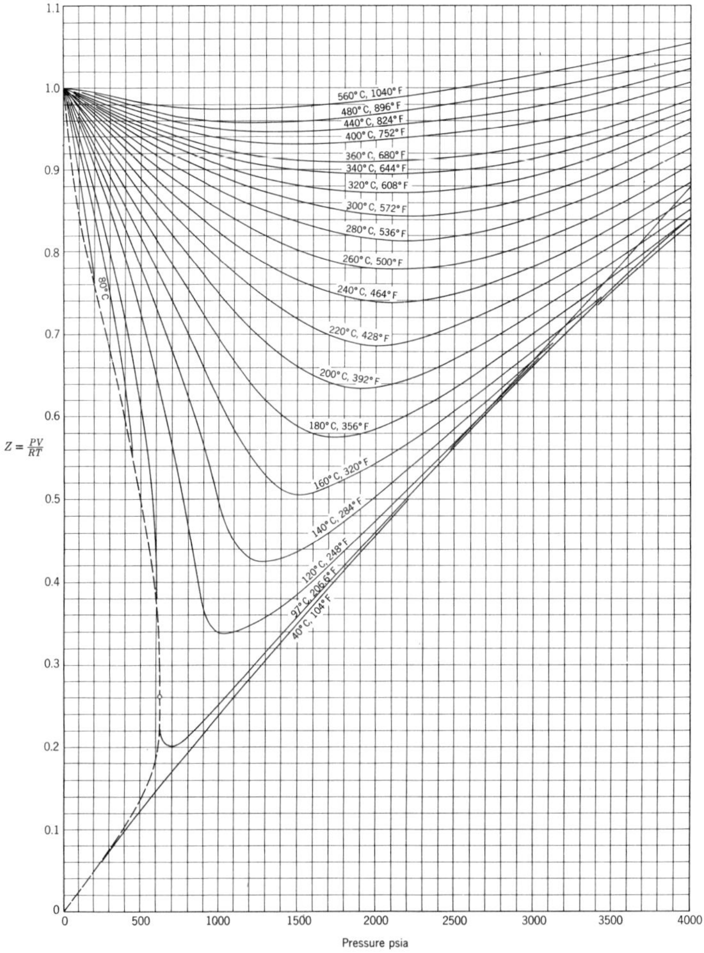

Fig. 8. Compressibility factors for propane. (Brown, Katz, Oberfell, and Alden, *Natural Gasoline and Volatile Hydrocarbons*, 1948, p. 27.)

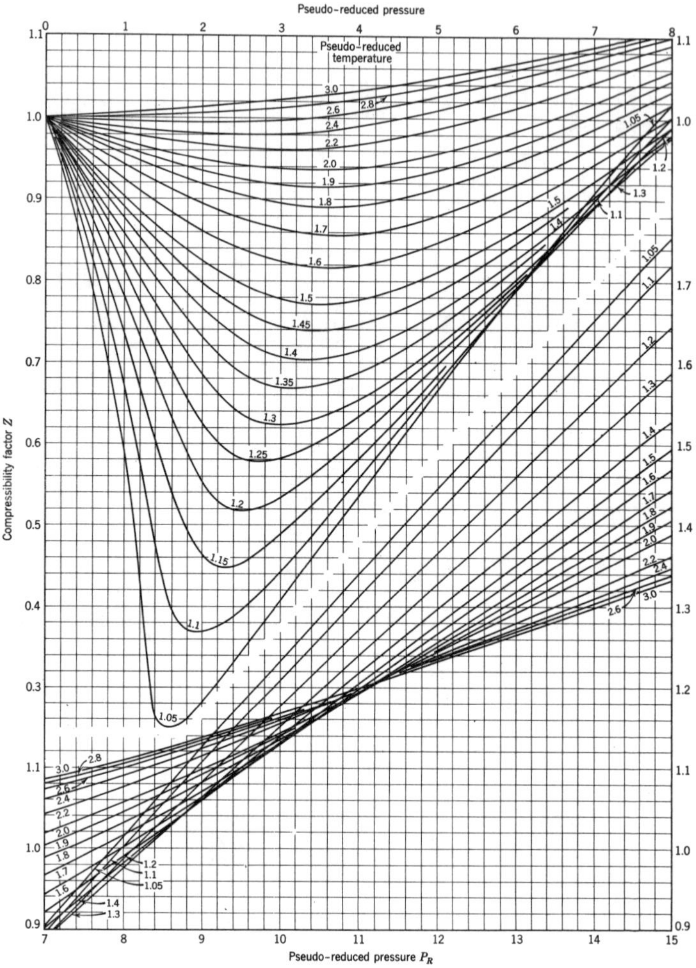

Fig. 10. *Z* factor for natural gases. (Standing and Katz, AIME Trans., 1942, p. 144.)

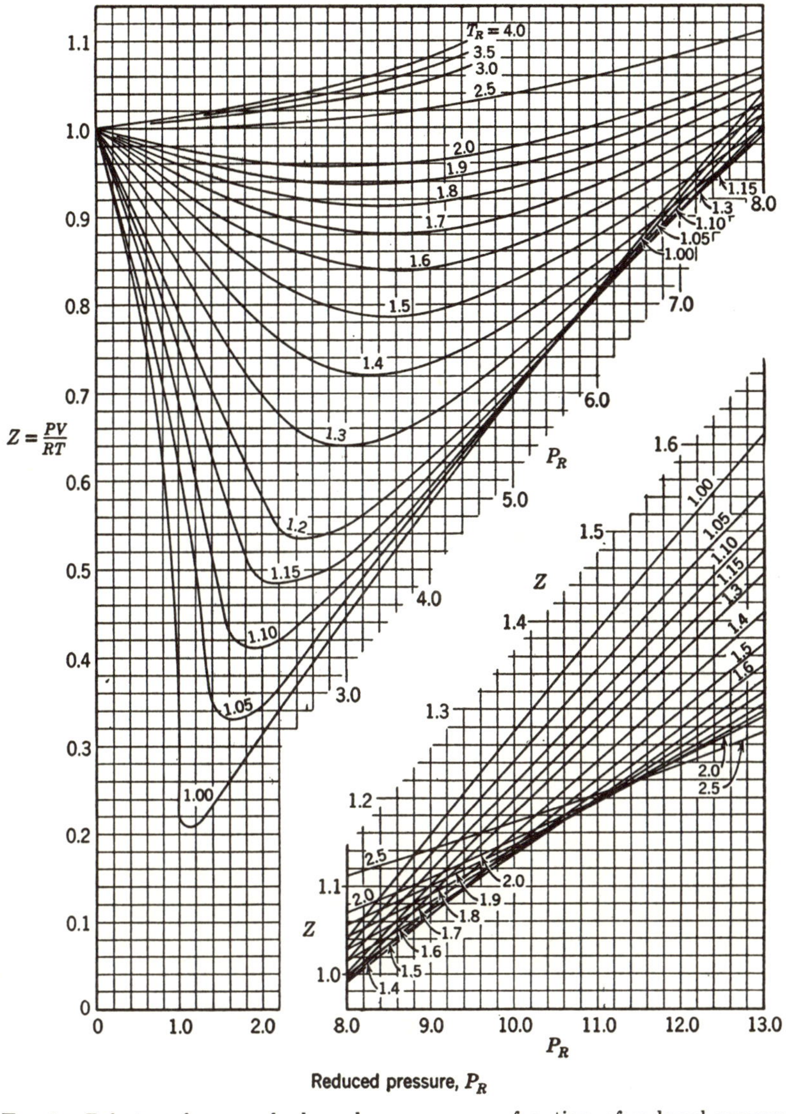

FIG. 9. Z factors for pure hydrocarbon gases as a function of reduced pressure and temperature. (Brown, Katz, Oberfell, and Alden, *Natural Gasoline and Volatile Hydrocarbons, 1948,* p. 32.)

EXAMPLE. A gas consists of 16 lb of methane and 7½ lb of ethane. Calculate Z at 0° F and 1000 psia.

This mixture consists of one mole of methane and 0.25 mole of ethane. Consequently, the mole fractions of methane and ethane are 0.80 and 0.20, respectively. Since P_c for methane is 673 and for ethane is 712, the pseudo-critical pressure is

$$\underline{P_c} = 0.8 \times 673 + 0.2 \times 712 = 680.8 \text{ psia}$$

Similarly, since T_c is 344° R for methane and 549° R for ethane the pseudo-critical temperature is

$$\underline{T_c} = 0.8 \times 344 + 0.2 \times 549 = 385.0° \text{ R}$$

Consequently, the pseudo-reduced pressure and temperature are

$$P_R = \frac{P}{\underline{P_c}} = \frac{1000}{680.8} = 1.47 \quad \text{and} \quad T_R = \frac{T}{\underline{T_c}} = \frac{460}{385} = 1.19$$

From Figure 10 the value of Z is found to be 0.645.

The chart in Figure 10 was prepared from experimental data on binary hydrocarbon gas mixtures and several natural gases covering a wide range in composition. The accuracy of this chart was checked with data from 13 natural gas samples of known composition. The Z factors of these 13 gases were determined experimentally at various temperatures and pressures and compared with the calculated values. The maximum deviation was found to be 6.7% and the average deviation was zero. This agreement between theory and experiment indicates that the method of calculating the compressibility factor presented above is sufficiently accurate for most engineering computations.

However, since this method is based on the Law of Corresponding States, it should not be applied with confidence to gases which contain relatively large amounts of non-hydrocarbon gases. It has been found that 4% carbon dioxide in a natural gas will cause errors of about 5% in the Z factor. However, considerably more nitrogen may be present without causing excessive error. The Z factor of gases containing as much as 20% nitrogen may be calculated by this method with an error of about 4%. Indeed, in the construction of the chart shown in Figure 10 some of the data used were from natural gas samples which contained as much as 4% nitrogen.

To apply the above method to mixtures, it is necessary to know the gas composition in order to calculate the pseudo-criticals. If data on the composition are not available, it is still possible to approximate a value for Z if the specific gravity of the gas is known. It has been found that if the pseudo-critical pressure is plotted as a function of specific gravity for a large number of gases, most points fall on a straight line. Similarly, a plot of pseudo-critical temperature versus specific gravity results in a smooth curve. These correlations are shown in Figures 11 and 12, and from these charts the pseudo-criticals can be estimated and the pseudo-reduced temperature and pressure calculated in the usual manner. It should be emphasized that this

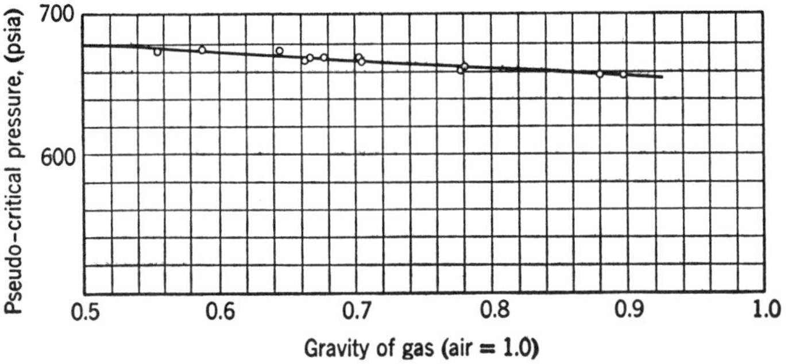

Fig. 11. Approximate pseudo-critical pressure as a function of gas gravity. (Natural Gasoline Supply Men's Association, *Engineering Data Book,* 1951, p. 82.)

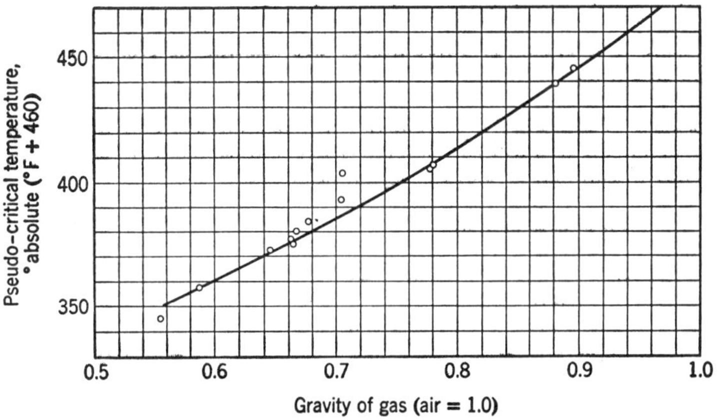

Fig. 12. Approximate pseudo-critical temperature versus gas gravity. (Natural Gasoline Supply Men's Association, *Engineering Data Book,* 1951, p. 82.)

method is not as accurate as the preceding method and should not be employed if the composition of the gas mixture is known.

Example. Estimate the Z factor for a 0.800 specific gravity gas at 1390 psia and 98° F. From Figures 11 and 12 the pseudo-critical pressure and temperature are found to be 662 psia and 413° R respectively. The pseudo-reduced pressure and temperature are

$$P_R = \frac{1390}{662} = 2.1 \qquad \text{and} \qquad T_R = \frac{460 + 98}{413} = 1.35$$

The Z factor is obtained from Figure 10 and found to be 0.72.

The use of compressibility factors to correct for the non-perfect behavior of gases is well suited for engineering calculations which usually require the computation of the volume occupied by a gas at a given temperature and pressure. However, it is sometimes necessary to calculate the pressure exerted by a non-perfect gas at a given volume and temperature or the temperature of a non-perfect gas at a given pressure and volume. Calculations of this type using a Z factor can best be illustrated by examples.

Example. Calculate the pressure of 15 lb of ethane in a one cu ft tank at a temperature of 140° F.

$$Z = \frac{PV}{nRT} = \frac{P \times 1}{0.5 \times 10.72 \times (460 + 140)} = 0.00031P$$

If this equation is plotted on Figure 7 a straight line, through the origin, results whose slope is 0.00031. The required pressure is given by the point of intersection of this straight line and the curve of Z versus P at a temperature of 140° F. The required pressure in this case would be 1265 psia.

If the gas is a mixture so that reduced temperature and pressure are being employed the calculations are similar as shown in the following example.

Example. Calculate the pressure of 0.333 moles of a gas mixture in a one cu ft tank at a temperature of 190° F if the pseudo-critical temperature and pressure of the mixture are 500° R and 700 psia, respectively.

$$T_R = \tfrac{650}{500} = 1.30$$

$$P_R = \frac{P}{700} \quad \text{or} \quad P = 700P_R$$

$$Z = \frac{PV}{nRT} = \frac{700P_R \times 1}{0.333 \times 10.72 \times (460 + 190)} = 0.302P_R$$

When $Z = 0.302P_R$ is plotted on Figure 10 this straight line intersects Z versus P_R for $T_R = 1.30$ at $P_R = 2.2$. Consequently, $P = 2.2 \times 700 = 1540$ psia.

It is apparent that a similar method of solution can be employed to calculate the temperature of a gas at a given pressure and volume.

REFERENCES

Brown, G. G., D. L. Katz, G. G. Oberfell, and R. C. Alden, *Natural Gasoline and the Volatile Hydrocarbons,* Natural Gasoline Association of America, Tulsa, Okla. (1948).

Calhoun, J. C., *Engineering Fundamentals,* University of Oklahoma Press, Norman, Okla. (1953).

Daniels, F., *Outlines of Physical Chemistry,* John Wiley & Sons, New York (1948).

Muskat, M., *Physical Principles of Oil Production,* McGraw-Hill Book Co., New York (1949).
Natural Gasoline Supplymen's Association Engineering Data Book.
Pirson, S., *Elements of Oil Reservoir Engineering,* McGraw-Hill Book Co., New York (1950).

PROBLEMS

1. If volume is to be measured in cubic feet, pressure in standard atmospheres, temperature in degrees Kelvin, and moles in lb moles, evaluate the gas constant for this system of units.

2. A single gaseous hydrocarbon has a density of 2.550 grams per liter at 100° C and one atmosphere. Chemical analysis shows that for each carbon atom in the molecule there is one atom of hydrogen. What is the formula of this hydrocarbon? *Answer:* C_6H_6.

3. The coefficient of thermal expansion is defined as the rate of change of volume with temperature at constant pressure per unit volume or

$$\text{Coefficient of thermal expansion} = \frac{1}{V}\left(\frac{\partial V}{\partial T}\right)_P$$

Evaluate this coefficient at 0° C for a perfect gas.

4. A 20 cu ft tank of ethane is evacuated until a pressure of 0.1 psia is obtained. If the temperature is 60° F., what weight of ethane remains in the tank? What is the gas density? *Answer:* 0.0108 lb, 0.00054 lb/cu ft.

5. How many pounds of methane are required to fill a tank of 20 cu ft capacity to a pressure of 2 atmospheres at 100° F? What would be the pressure if the temperature were reduced to 50° F?

6. Show that the apparent molecular weight of air is 28.96 using the composition given in Table 3. The weight of one liter of dry air is 1.2745 grams at 0° C and 14.5 psia. Calculate the apparent molecular weight of air using this data.

7. A gas has the following composition:

Component	Mole Fraction
Methane	0.890
Ethane	0.050
Propane	0.020
Isobutane	0.010
Butane	0.030

What are the weight fractions, the apparent molecular weight, the specific gravity? If the total pressure is 20 psia, what is the partial pressure of each gas?

8. A gas has the following composition:

Component	Weight Fraction
Methane	0.700
Ethane	0.070
Propane	0.100
Butane	0.100
Pentane	0.030

What is the composition in mole %?

9. A gas field which is 25 sq miles in area produces from a sand which has an average thickness of 50 ft. The temperature and pressure of the reservoir are 180° F and 625 psi gage. The porosity of the sand is 19.0%. The gas has the following composition:

Component	Volume Fraction
Methane	87.2%
Ethane	9.4%
Propane	0.6%
Carbon Dioxide	1.8%
Nitrogen	1.0%

Calculate, assuming perfect gas behavior:

(*a*) The number of standard cubic feet of gas which can be produced from the reservoir.

(*b*) The number of pounds of gas originally in place. *Answer:* 11.2×10^9 lb.

10. Two cylinders of equal volume contain methane and ethane at 125 psia and 50 psia, respectively, at 100° F. If they are connected together and the gases allowed to mix, what are the final pressure, the partial pressures, and the composition of the resulting mixture?

11. A natural gas consists of 90% by volume of methane and ethane, and 10% by volume of propane. If the specific gravity of the gas is 0.75, what percentages of methane and ethane are in the gas by volume, by weight, and by moles?

12. A 10 cu ft tank contains a single paraffin gas at 150 psia and 95° F. When 5 lb of methane are added the specific gravity of the gas mixture is 0.983. What gas was originally present in the tank? *Answer:* Propane.

13. A centimeter cube of a sandstone is placed in an air-filled vessel whose volume is 10.00 cm^3. The pressure is 750 mm of mercury. The vessel is sealed and a stopcock turned so that the air can expand into a second evacuated vessel whose volume is also 10.00 cm^3. The final pressure in the two vessels is 361.4 mm of mercury. Calculate the porosity of the sandstone.

14. Calculate Van der Waals' constants for propylene. ($P_c = 45.4$ atm, $V_c =$ 180 cc, and $T_c = 91.4°$ C). Calculate the pressure exerted by one mole of propylene at 100° C in a vessel having a volume of 20.06 liters. Calculate the pressure assuming perfect gas behavior. Repeat these calculations if the volume is 0.20 liter.

15. One mole of methane is contained in a vessel under a pressure of 333 psia at −47° F. Assuming perfect gas behavior calculate the volume. Assuming $PV = ZnRT$ calculate the volume, using both Figures 6 and 9 to obtain the Z factor, and compare the results.

16. Compute the pseudo-critical temperatures and pressures for the gas mixtures in Problems 7 and 8.

17. A gas has the following composition:

Component	Mole Fraction
CO_2	0.0060
CH_4	0.8811
C_2H_6	0.0601
C_3H_8	0.0506
Iso C_4H_{10}	0.0011
C_4H_{10}	0.0011

Compute Z at 235° F and 1000 psia and at 100° F and 2000 psia (use Figure 10).

18. Calculate the specific gravity of the gas in Problem 17. Now calculate Z at the same temperatures and pressures, using the curves in Figures 11 and 12.

19. Work problem 9 using a Z factor obtained by each of the two methods studied.

20. A 10 cu ft tank at a pressure of 2000 psia contains ethane gas. If the Z factor is 0.60, how many pounds of ethane are in the tank? *Answer:* 134 lb.

21. A non-ideal gas consists of 0.25 moles of methane and 0.75 moles of propane. Calculate the pressure exerted by this gas if it exists in a 2 cu ft tank at a temperature of 215° F.

CHAPTER 3

PHASE BEHAVIOR OF LIQUIDS

It is difficult to state a concise definition of a liquid in a manner that clearly distinguishes it from a gas. This is due to the fact that the liquid and the gaseous states are in reality closely related (see Chapter 4), and it is possible to pass continuously from one to the other without any abrupt change in state. Generally speaking, however, liquids are homogeneous fluids that have higher densities than gases. Furthermore, the molecules are more closely packed in a liquid than in a gas, so that the attractive forces between liquid molecules are appreciable. These intermolecular forces are of sufficient magnitude to maintain a liquid in one mass when it is placed in a vessel whose volume is larger than that of the liquid. However, the molecules still possess sufficient freedom to allow the liquid to flow, although the viscosities of liquids are generally greater than those of gases.

Pressure, Volume, Temperature Relations for a Liquid. In Chapter 2 it was shown that the behavior of gases could be expressed by a few simple laws which are, to a large extent, independent of the nature of the gas. This is not true of liquids, and no simple law, analogous to the general gas law for gases, can be written governing the behavior of liquids. To be sure, expressions for the volume of a liquid as a function of temperature and pressure can be written. At constant pressure the volume of a liquid at any temperature T is given by

$$V = V_0[1 + \alpha(T - T_0)] \tag{1}$$

where V_0 is the initial liquid volume at T_0 and α is the average coefficient of thermal expansion in the temperature interval from T_0 to T. However, the value of α, at a given temperature and pressure, will be different for various liquids. Similarly, at constant temperature, the following expression can be written for the volume of a liquid at pressure P

$$V = V_0[1 - C(P - P_0)] \tag{2}$$

where V_0 is the initial volume at P_0 and C is the average coefficient of compression for the pressure interval P_0 to P. Here again, each liquid has a different value for C. The values of α and C are generally small numbers when compared to the corresponding quantities for gases, but it is obvious that a prediction of the volume variations of a liquid due to changes in pressure and temperature requires a knowledge of α and C for the particular liquid in question.

The Vapor Pressure of Liquids. Vapor pressure is defined as the pressure exerted by a vapor in equilibrium with its liquid. Consider a closed, evacuated container which has been partially filled with a liquid. The molecules of the liquid are in constant motion but not all the molecules move with the same velocity and there will be some which possess a relatively high kinetic energy. If one of these fast moving molecules reaches the liquid surface, it may possess sufficient energy to overcome the attractive forces in the liquid and pass into the vapor space above. As the number of molecules in the vapor phase increases the rate of return to the liquid phase also increases and eventually a condition of dynamic equilibrium is attained when the number of molecules leaving the liquid is equal to the number returning. The molecules in the vapor phase obviously exert a pressure on the containing vessel and this pressure is known as the vapor pressure.

Vapor Pressure as a Function of Temperature. As the temperature of a liquid increases, the average molecular velocity increases and con-

Table 6. Vapor Pressures of Water and *n*-Hexane

Temperature (° C)	Vapor Pressure of Water (mm of Hg)	Vapor Pressure of *n*-Hexane (mm of Hg)
0	4.58	45.4
20	17.54	120.0
40	55.30	276.7
60	149.4	566.2
69	.....	760.0
80	355.1	1062
100	760.0	1836

sequently a larger number of molecules possess sufficient energy to enter the vapor phase. As a result the vapor pressure of a liquid increases with increasing temperature. This increase in vapor pressure with temperature is shown for water and n-hexane in Table 6.

Figure 13 shows an illustrative plot of vapor pressure versus temperature. It is apparent from this figure that the vapor pressure is not a linear function of temperature. However, by a suitable choice of coordinate axes a linear relationship between vapor pressure and

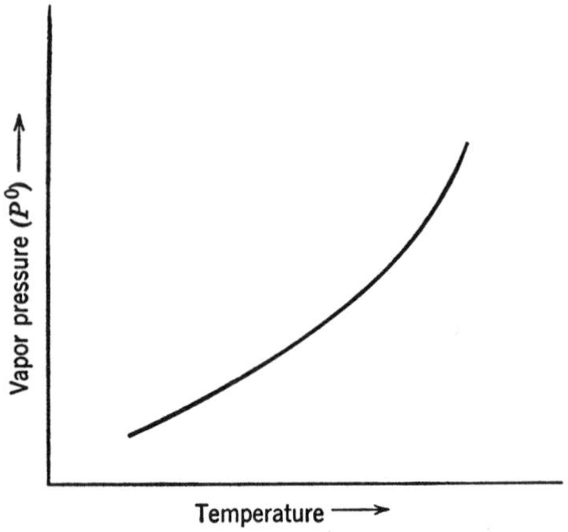

FIG. 13. Illustrative plot of vapor pressure as a function of temperature.

temperature can be obtained. A method which is particularly convenient for plotting vapor pressure data for hydrocarbons is shown in Figure 14. This is known as a Cox chart. The vapor pressure scale is logarithmic but the temperature scale in ° F is entirely arbitrary. A straight line is drawn on the chart and vapor pressure data for water is used in conjunction with this straight line to determine the temperature scale. It has been found that the majority of the hydrocarbons that commonly occur in petroleum give straight lines when their vapor pressures are plotted on this chart. Furthermore, the straight lines obtained in this manner for the petroleum hydrocarbons appear to have a common point of intersection (off the chart in the upper right hand corner of Figure 14). Consequently, if the vapor pressure of an unknown hydrocarbon is known at some temperature, its vapor pressure at other temperatures can be estimated by a straight line drawn through the known point and the common point of intersection.

As shown in Figure 15, a straight line is also obtained when the logarithm of the vapor pressure is plotted as a function of the recip-

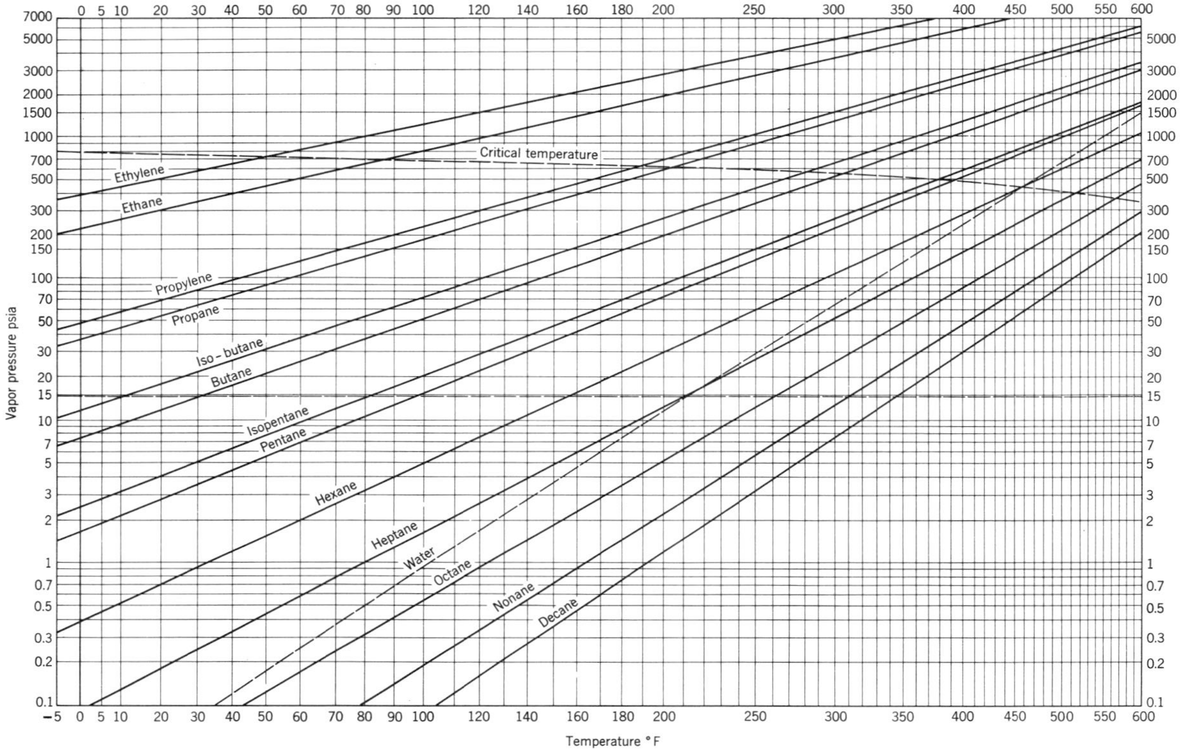

Fig. 14. Vapor pressure of various hydrocarbons as a function of temperature. (Rector, Natural Gasoline Supply Men's Association, *Engineering Data Book*, 1951, pp. 104, 105.)

rocal of the *absolute* temperature. The reason for this behavior will be explained in a later section of this chapter.

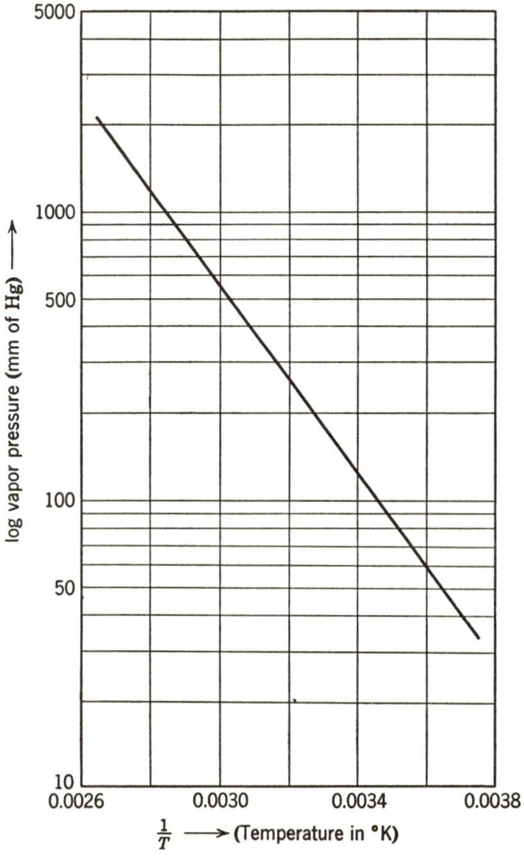

FIG. 15. Logarithm vapor pressure versus reciprocal of the absolute temperature. Data for *n*-Hexane.

Measurement of Vapor Pressure. A number of experimental methods for the measurement of vapor pressure are available. Two of these methods will be briefly described since they illustrate the physical meaning of vapor pressure.

The first method is a direct method and employs an apparatus which is diagrammatically represented in Figure 16. The entire apparatus is enclosed in a constant temperature bath. The liquid whose vapor pressure is to be determined is placed in vessel *A* and the system is evacuated to remove all air from both sides of the mercury manometer. When all the air has been removed, the two stopcocks *B* and *C* are closed and the system is allowed to come to thermal equilibrium. The vapor pressure of the liquid at the temperature of the bath is read directly on the mercury manometer. Obviously, the vapor pressure at other temperatures can be determined simply by altering the bath temperature.

A second method for the determination of vapor pressure which is based on the general gas law is known as the "Gas Saturation" method. A weighed quantity of a liquid of known molecular weight is placed

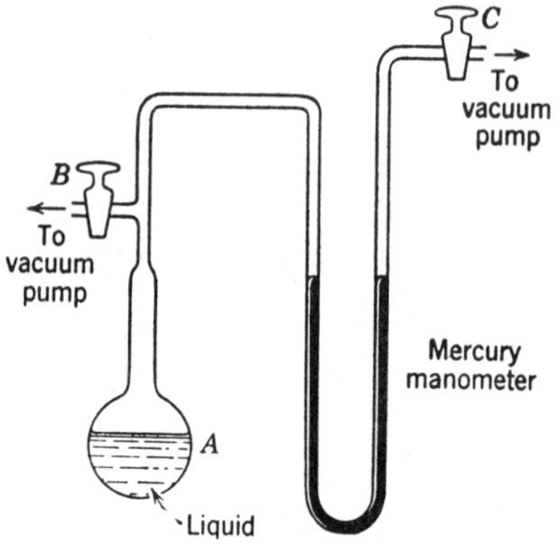

FIG. 16. Diagrammatic representation of an apparatus for measuring vapor pressure of a liquid.

in a glass trap (Figure 17) and a known volume of air is passed in and bubbled through the liquid. The loss in weight, due to evaporation of the liquid, is determined by direct weighing. This weight of liquid existing in the vapor state can be considered to occupy a volume very nearly equal to the volume of the air passed into the liquid. Applying the general gas law, the partial pressure of the vapor which is equal to the vapor pressure can be calculated using the following equation:

$$\text{Vapor pressure} = P^\circ = \frac{\text{Wt} \times RT}{\text{MW} \times V} \tag{3}$$

where Wt = weight of liquid vaporized (weight of vapor), R is the gas constant, T is the absolute temperature at which the experiment is carried out, MW is the molecular weight of the vapor and V is the volume of air passed into the liquid. Obviously, this vapor pressure is for the temperature T.

Equation 3 is not exact since V in this equation should be the volume of air and vapor that comes out of the trap and not the volume of air passed in. The higher the vapor pressure of the liquid, the greater the difference between the volume of air (V_1) which is passed into the trap and the volume which comes out (V_2). Consequently, if the vapor pressure is small, this simple method will give acceptable results. However, if the vapor pressure is large a correction must be applied. This can be done by applying Boyle's Law to the *air*. If P

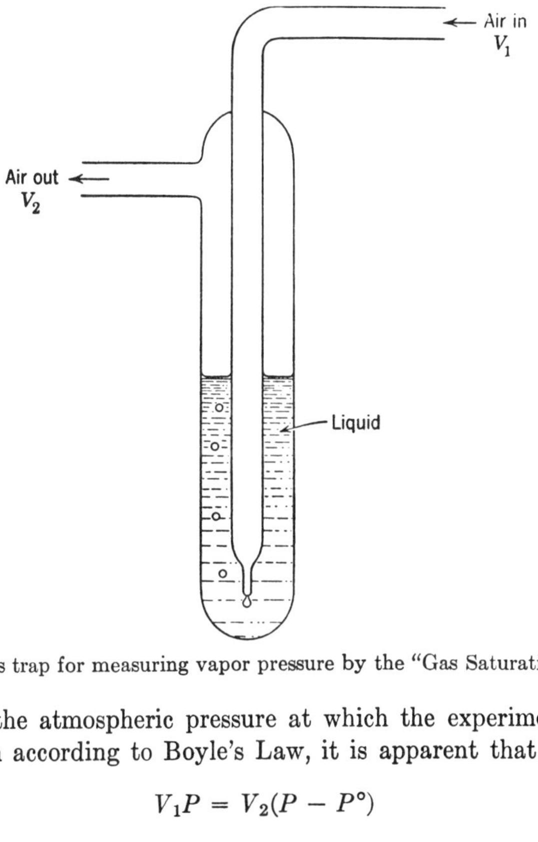

FIG. 17. Glass trap for measuring vapor pressure by the "Gas Saturation" method.

represents the atmospheric pressure at which the experiment is conducted then according to Boyle's Law, it is apparent that

$$V_1P = V_2(P - P°)$$

therefore

$$V_2 = \frac{V_1P}{P - P°}$$

Substituting this value of V_2 into equation 3 for V, it follows that

$$P° = \frac{\text{Wt} \times RT}{(\text{MW} \times V_1P)/(P - P°)}$$

Solving for $P°$ gives

$$\text{Vapor pressure} = P° = \frac{(\text{Wt} \times RT)/(\text{MW} \times V_1)}{1 + (\text{Wt} \times RT)/(\text{MW} \times V_1P)} \qquad (4)$$

This more exact equation requires a knowledge of the atmospheric pressure in addition to the experimental data necessary to calculate an approximate vapor pressure by equation 3.

To obtain a reliable value of the vapor pressure by this method it is necessary that the air is completely saturated with the vapor. To bring this about the air is passed through the liquid at a slow rate and two, or even three, traps are placed in series. If the weight of the last trap remains unchanged during the course of an experiment, one can be certain that the air was completely saturated in passing through the preceding traps.

The Clausius–Clapeyron Equation. This equation expresses the relationship between vapor pressure and temperature. In 1834 Clapeyron, using thermodynamic theory, developed the following equation which will be accepted without derivation

$$\frac{dP^\circ}{dT} = \frac{\Delta H}{T\Delta V}$$

where dP°/dT is the rate of change of vapor pressure with temperature, ΔH is the heat of vaporization of a given quantity of liquid, T is the absolute temperature, and ΔV represents the change in volume of the given quantity of liquid in going from the liquid to the gaseous state (for derivation see, for example, Daniels, *Outlines of Physical Chemistry*). If it is assumed that the vapor is a perfect gas the Clapeyron equation can be simplified in the following manner.

Consider one mole of liquid. Consequently, ΔH will be the heat of vaporization of one mole (ΔH_m) and ΔV will be given by

$$\Delta V = V_g - V_l$$

where V_g and V_l represent the volumes of one mole of gas and one mole of liquid respectively. Since the liquid molar volume is very much smaller than the molar volume of the gas it can be neglected for all practical purposes. Under these conditions the Clapeyron equation becomes

$$\frac{dP^\circ}{dT} = \frac{\Delta H_m}{TV_g} \tag{5}$$

If the vapor is a perfect gas, it is apparent that

$$V_g = \frac{RT}{P^\circ}$$

Substitution for V_g in equation 5 leads to

$$\frac{dP^\circ}{dT} = \frac{\Delta H_m P^\circ}{RT^2} \quad \text{or} \quad \frac{d \ln P^\circ}{dT} = \frac{\Delta H_m}{RT^2}$$

If it is assumed that ΔH_m is a constant independent of temperature, this equation can be integrated to give

$$\ln P^\circ = -\frac{\Delta H_m}{RT} + C$$

where C is the constant of integration. This equation clearly expresses the fact that the logarithm of vapor pressure is a linear function of $1/T$ (see Figure 15) whose slope is $-\Delta H_m/R$. If the integration is carried out between limits, the following equation results

$$\ln \frac{P^\circ_2}{P^\circ_1} = \frac{\Delta H_m}{R}\left(\frac{1}{T_1} - \frac{1}{T_2}\right) \tag{6}$$

This is known as the Clausius–Clapeyron equation. If the molar heat of vaporization and the vapor pressure at some temperature are known for a liquid, the vapor pressure at other temperatures can be calculated, provided the assumptions made in the derivation of this equation are valid. Since the normal boiling point of a liquid is defined as the temperature at which the vapor pressure equals one atmosphere, it is apparent that only the molar heat of vaporization and the normal boiling point of a liquid need to be known in order to calculate the vapor pressure at other temperatures.

Since ΔH_m in equation 6 is expresed in calories per gram-mole or in Btu per pound-mole, it is apparent that R must be expressed in calories per degree per gram-mole or in Btu per degree per pound-mole, respectively. It can be shown that the value of R in both systems of units is equal to 1.987. Since the vapor pressures occur as a ratio in this equation their units are unessential, provided they are the same. An absolute temperature scale must be used however. If ΔH_m is in Btu per pound-mole then T must be in ° R. If ΔH_m is in calories per gram-mole then T must be in ° K. An example calculation which illustrates the use of the Clausius–Clapeyron equation is given below.

Example. The molal heat of vaporization of a hydrocarbon is 5360 calories. The vapor pressure at 7° C is 1000 mm of mercury. What is the vapor pressure in psia at 20° C?

Let $P^\circ_1 = 1000$ mm, $T_1 = 280°$ K, $T_2 = 293°$ K.

$$\Delta H_m = 5360 \text{ calories} \qquad R = 1.987$$

$$\ln \frac{P^\circ_2}{1000} = \frac{5360}{1.987}\left(\frac{1}{280} - \frac{1}{293}\right)$$

$$= 0.427$$

Therefore $$\frac{P^\circ_2}{1000} = 1.53 \quad \text{or} \quad P^\circ_2 = 1530 \text{ mm}$$

1530 mm = 29.6 psia = the required vapor pressure at 20° C.

The Heat of Vaporization. The molal heat of vaporization represents the energy that must be supplied to vaporize one mole of the liquid. For example, 9714 calories must be supplied to vaporize one mole of water at 100° C. The heat of vaporization does depend upon the temperature at which the vaporization is carried out. However, the variation is in general not great, so that the assumption made in the previous section regarding the constancy of ΔH_m is justifiable over a small temperature range.

The molar heats of vaporization of several hydrocarbons are given in Table 7. The normal boiling points in ° F are also given. The

Table 7. Normal Boiling Points and Heats of Vaporization of Various Hydrocarbons

Hydrocarbon	Heat of Vaporization (Btu/lb-mole)	Boiling Point (at 1 atm, ° F)	$\frac{\Delta H_m}{T}$ (T = ° R)
CH_4	3,930	−258.5	19.5
C_2H_6	6,344	−128.2	19.1
C_3H_8	8,069	−43.8	19.4
i-C_4H_{10}	9,183	10.9	19.5
C_4H_{10}	9,648	31.1	19.6
i-C_5H_{12}	10,533	82.2	19.4
C_5H_{12}	11,038	97.0	19.8
C_6H_{14}	12,581	155.7	20.4
C_7H_{16}	13,827	209.1	20.7
C_8H_{18}	14,963	258.1	20.8
C_9H_{20}	16,031	303.3	21.0
$C_{10}H_{22}$	17,073	345.2	21.2
		Average:	20.0

values of $\Delta H_m/T$, where T is the normal boiling point in ° R, are also included in this table. It is apparent that they are remarkably constant for all the hydrocarbons listed. This is known as Trouton's Rule. If the molar heat of vaporization is expressed in calories and the boiling point in degrees Kelvin, Trouton's constant has the same value. The literature generally gives 21 as the value of Trouton's constant.

This is the average value of $\Delta H_m/T$ for a large number of liquids of all types. For the hydrocarbons 20 is apparently a better average value. An obvious application of Trouton's Rule would be the calculation of an approximate vapor pressure for a liquid whose normal boiling point is known but for which heat of vaporization data are unavailable.

REFERENCES

Calingaert and Davis, *Ind. and Eng. Chem. 17,* 1287 (1925).

Cox, *Ind. and Eng. Chem. 15,* 592 (1923).

Daniels, F., *Outlines of Physical Chemistry,* John Wiley & Sons., New York (1948).

Prutton and Marion, *Fundamental Principles of Physical Chemistry,* The Macmillan Company, New York (1949).

PROBLEMS

1. Plot vapor pressure of pentane as a function of temperature from 60° to 200° F. Plot log vapor pressure versus $1/T$ over the same temperature range.

2. Construct a Cox chart for *n*-hexane using the vapor pressure data given in Table 6. Use vapor pressure data for water to construct the arbitrary temperature scale.

3. A hydrocarbon has the following vapor pressures

Temperature	Vapor Pressure
−75° F	6.36 psia
−50° F	12.60 psia
−25° F	22.70 psia

Calculate graphically by plotting logarithm of the vapor pressure versus $1/T$; (*a*) Vapor pressure at 40° F. *Answer:* 78 psia; (*b*) Boiling point at one standard atmosphere pressure; (*c*) The gage storage pressure required to prevent loss by evaporation at 0° F.

4. 20 liters of dry air is passed through a pure liquid hydrocarbon (MW = 144) at 20° C. The loss in weight of the liquid was 1.310 grams. What is the vapor pressure at this temperature? If the atmospheric pressure is 750 mm, calculate a more accurate vapor pressure. *Answer:* 0.0109 atm, 0.0108 atm.

5. The heat of vaporization of ether (MW = 46) is 88.4 calories per gram at its normal boiling point (34.5° C). Calculate the vapor pressure at 60° C. At what temperature is the vapor pressure equal to 280 mm?

6. The vapor pressure of a pure hydrocarbon is 6.36 psia at −75° F. If the heat of vaporization is 8450 Btu per pound-mole, what is the vapor pressure at 40° F?

7. When 20 liters of air measured at 760 mm are passed through a liquid at 20° C the weight of liquid evaporated is 0.343 grams. If the molecular weight of the liquid is 18 and the heat of vaporization is 540 calories per gram, what is the vapor pressure at 90° C?

8. A pure hydrocarbon liquid boils at 100° F under atmospheric pressure. What is an approximate value of the vapor pressure at 150° F? *Answer:* 33.6 psia.

CHAPTER 4

QUALITATIVE PHASE BEHAVIOR OF HYDROCARBON SYSTEMS

The properties of gases and liquids were described in Chapters 2 and 3. This chapter will treat the behavior of systems that are made up of matter in two or more states of aggregation. These heterogeneous systems are said to consist of two or more *phases,* a phase being defined as a physically distinct portion of matter having uniform physical and chemical properties throughout. Thus, a system composed of ice, water, and water vapor is said to be a three-phase system. Although a phase is homogeneous, it need not necessarily be continuous. An ice-water system, for example, consists of *two* phases regardless of the state of subdivision of either the ice or the water.

The properties of a phase are either intensive or extensive. An intensive property is one which is independent of the total quantity of matter in the system. Examples are density, specific gravity, and specific heat. Properties such as the mass and volume of a system are termed extensive properties since their value is determined by the quantity of matter contained in the system.

It will be shown that the behavior of heterogeneous systems is influenced by the number of components it contains. A system which consists of a single, pure substance will behave differently from one which is made up of two or more components when the pressure and temperature are such that both a liquid phase and a gas phase are present. Consequently, the discussion of phase behavior will begin with a description of single-component systems. This will be followed by a description of two-component systems. Finally, multicomponent systems will be considered. In this chapter the *qualitative* behavior of these systems will be of primary interest and the quantitative treatment will be presented in the following chapter.

SINGLE-COMPONENT SYSTEMS

Consider a single, pure fluid at constant temperature, in a cylinder fitted with a frictionless piston. If a pressure is applied on the piston which is *greater* than the vapor pressure of the liquid, the system will consist entirely of liquid when equilibrium is reached. No vapor will be present since at pressures greater than the vapor pressure it condenses into liquid. If, on the other hand, the pressure applied on the piston is *less* than the vapor pressure of the liquid only vapor will be present at equilibrium. If *both* liquid and vapor are present in equilibrium with one another, the pressure must be *exactly equal* to the vapor pressure. *Pure substances behave in this manner and liquid and vapor can coexist at a given temperature only at a pressure equal to the vapor pressure.* The relative amounts of liquid and vapor that coexist is determined by the volume of the system, and can vary anywhere from an infinitesimal amount of liquid to an infinitesimal amount of vapor.

The Pressure-Temperature Diagram for a Pure Substance. For a single-component system at a given temperature the pressure determines the kind and number of phases that are present. If the vapor pressure is plotted as a function of temperature, the resulting curve can be thought of as being the dividing line between the area where liquid exists and the area where vapor or gas exists. If the line *OA* (Figure 18) represents the vapor pressure as a function of temperature the systems which are represented by points above *OA* are composed of liquid only. Similarly, points below *OA* represent systems that are all vapor. If the system is represented by a point on the line *OA* then the system consists of both liquid and vapor.

The upper limit of the vapor pressure line is the point *A*. This is known as the critical point and the temperature and pressure represented by this point are the critical temperature T_c and the critical pressure P_c, respectively. At this point the intensive properties of the liquid phase and the vapor phase become identical and they are no longer distinguishable. For a single-component system the critical temperature may also be defined as the temperature above which a vapor cannot be liquefied, regardless of the applied pressure. Similarly, the critical pressure of a single-component system may be defined as the minimum pressure necessary for liquefaction of vapor at the critical temperature. It is also the pressure above which liquid and vapor cannot coexist regardless of the temperature.

The lower end of the vapor-pressure line is limited by the triple point O. This point represents the pressure and temperature at which solid, liquid, and vapor coexist under equilibrium conditions. Since the petroleum engineer seldom deals with hydrocarbons in the solid state it will not be necessary to deal with this region of the diagram

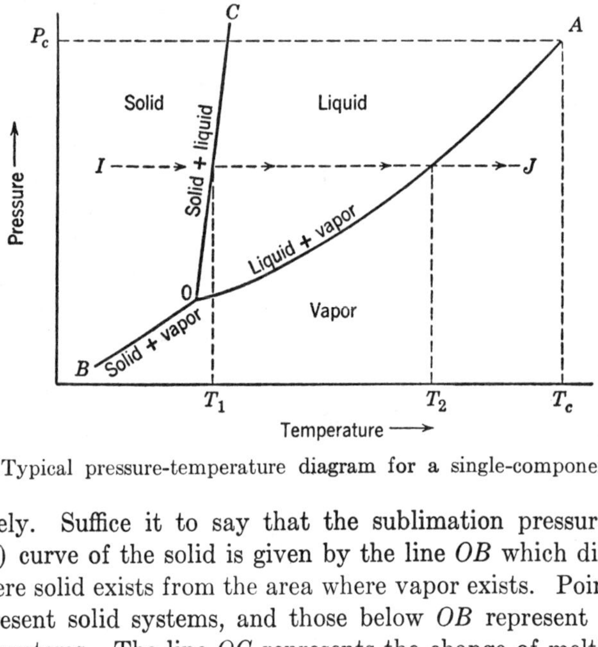

FIG. 18. Typical pressure-temperature diagram for a single-component system.

extensively. Suffice it to say that the sublimation pressure (vapor pressure) curve of the solid is given by the line OB which divides the area where solid exists from the area where vapor exists. Points above OB represent solid systems, and those below OB represent vapor or gaseous systems. The line OC represents the change of melting point with pressure and divides the solid area from the liquid area. For pure hydrocarbons the melting point generally increases with pressure so the slope of the line OC is positive as shown. Water is exceptional in that its melting point decreases with pressure so in this case the slope of the line OC is negative.

Each pure hydrocarbon has a pressure-temperature diagram similar to the one shown in Figure 18. To be sure, the actual vapor pressures, sublimation pressures, critical values, etc., are different for each substance, but the general characteristics are similar. If such a diagram is available for a given substance, it is obvious that it could be used to predict the behavior of the substance as the temperature and pressure are varied. For example, in the diagram shown in Figure 18, suppose the system is initially at a pressure and temperature represented by the point I and the system is heated at constant pressure until the point J is reached. For this isobaric temperature increase the following phase changes occur. The system is originally in the solid state and no phase change occurs until the temperature T_1 is reached. At this temperature, which is the melting point at this pressure, liquid will begin to form and the temperature will remain constant until all the solid has disappeared. As the temperature is further increased the system will be in the liquid state until the temperature T_2 is reached. At T_2, which is the boiling point at this pressure, vapor forms and again the temperature will remain constant until all the liquid has vaporized. The temperature of this vapor system can now be increased until the point J is reached. It should be emphasized that, in the process just described, only the *phase* changes were considered. For example, in going from just above T_1 to just below T_2 it was stated that only liquid was present and no phase change occurred. Obviously, the intensive properties of the liquid change as the temperature is increased. For instance, the increase in temperature causes an increase in volume with a resultant decrease in density. Similarly, other physical properties of the liquid are altered, but the properties of the system are those of a liquid and no other phases appear during this part of the isobaric temperature increase.

In order to obtain a better understanding of the phase behavior of liquids and vapors consider the region near the critical point in greater detail. Consider two systems whose initial temperature and pressure are represented by points A and B in Figure 19. If system A is heated at constant pressure a second, *less* dense phase will form at point D. Comparing the intensive properties of the two systems at D suggests that the original system at A was a liquid. Similarly, by cooling system B to D at constant pressure a second, *more* dense phase appears. This suggests that the original system at B was a vapor or gas. Now consider the following sequence of processes. Starting with the system in the liquid state at A increase the pressure isothermally to a value greater than P_c at point E. Keeping the pressure constant, increase the temperature to a value greater than T_c at point F. Now decrease the pressure to its original value at point G. Finally, decrease the temperature, keeping the pressure constant until point B is reached. The system is now in the vapor state and the transition from the liquid to the vapor state has been effected without an abrupt phase change. This shows that the liquid and vapor phases are in reality very similar. The vapor and liquid states are only separate forms of the same condition of matter, and it is possible to pass from one into the other by a series of gradations so small that there is never an abrupt phase change. Therefore, it is clear that the terms liquid and vapor on a phase diagram have a definite meaning only in the two-phase region. Consequently, in areas far removed from the two-phase region, particularly in the region where the pressure and tem-

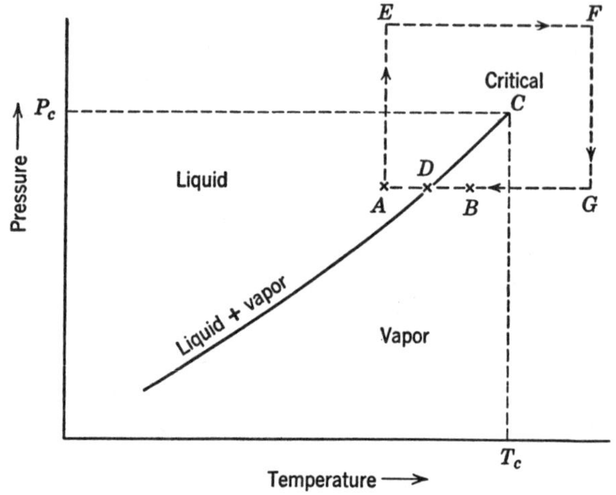

FIG. 19. Typical pressure-temperature diagram for a single-component system in the region near the critical point.

perature are above the criticals, definition of the liquid or gaseous states is impossible and the system is described as being in the *fluid* state.

In this discussion the terms vapor and gas have been used interchangeably but sometimes a distinction is made. The term vapor is sometimes applied to the less dense phase when it coexists with the more dense liquid phase or when its pressure and temperature are such that the point which represents this system on the pressure-temperature diagram is in the area immediately below the vapor pressure *versus* temperature line. The term gas is sometimes applied to systems which are represented by points far below the vapor pressure *versus* temperature line. Obviously this distinction is only relative.

The Pressure-Volume Diagram for a One-Component System. Another way of describing the phase behavior of a system is by means of a pressure-volume diagram. Pressure is plotted as a function of the volume and the behavior of the system at constant temperature is described. Consider a fixed quantity of a pure fluid at a fixed temperature whose initial pressure and volume are represented by the point *A* on the diagram in Figure 20. Furthermore, consider that this initial pressure is low enough so that the entire system is in the vapor or gaseous state. At constant temperature a decrease in volume is represented by the curve *AB*. As the volume decreases the pressure

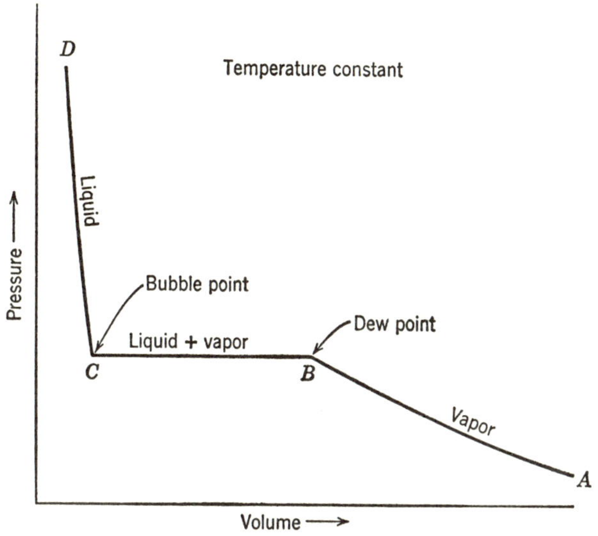

FIG. 20. Typical isotherm on a pressure-volume diagram for a single-component system.

increases and eventually becomes equal to the vapor pressure, provided the temperature is below the critical temperature. When this occurs, liquid begins to condense. This point is known as the dew point and is represented on the diagram by the point *B*. It has been shown previously that, for a single-component system at a constant temperature, liquid and vapor coexist at the vapor pressure. Consequently, the pressure remains constant as more and more liquid condenses and the volume of the system decreases. This process is represented by the *straight, horizontal* line *BC*. The point *C* is known as the bubble point and represents a system which is all liquid except for an infinitesimal amount of vapor. A characteristic of a single-component system is that at a given temperature, the vapor pressure, the dew-point pressure, and the bubble-point pressure are equal. Due to the fact that liquids are relatively incompressible a further decrease in volume can be brought about only by a relatively large pressure increase. Consequently, the isotherm *CD* is nearly perpendicular. This diagram represents a pressure-volume diagram for a pure substance. The line AB represents an isothermal traverse through the vapor region, the line BC a traverse through the two-phase region, and CD a traverse through the liquid region.

It is customary to include several isotherms on a pressure-volume diagram. Such a diagram is shown in Figure 21. The isotherm at the critical temperature T_c gives an inflection at the point C. C represents the critical point and P_c is the critical pressure. If the system

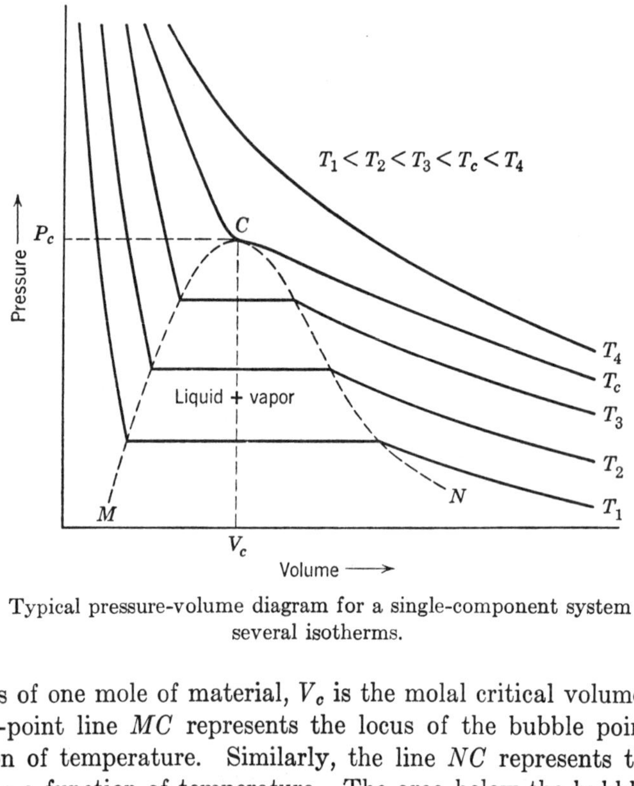

FIG. 21. Typical pressure-volume diagram for a single-component system showing several isotherms.

consists of one mole of material, V_c is the molal critical volume. The bubble-point line MC represents the locus of the bubble point as a function of temperature. Similarly, the line NC represents the dew point as a function of temperature. The area below the bubble-point line and the dew-point line represents the two-phase region.

It is of interest to consider the relationship between the pressure-volume diagram and the pressure-temperature diagram previously described. The straight, horizontal lines through the two-phase region on the P-V diagram represent the vapor pressures at the temperatures for which the isotherms are drawn. If these pressures are plotted as a function of temperature the line OA on the P-T diagram is obtained (Figure 18). Since the vapor pressure, dew-point pressure, and bubble-point pressure are equal at a given temperature, it is apparent that the line OA on the P-T diagram also represents the dew-point and bubble-point pressure as a function of temperature. In other words, for a single-component system, the vapor pressure line, the dew-point pressure line, and the bubble-point pressure line on a P-T diagram all coincide.

The Density-Temperature Diagram. Consider the densities of the liquid and vapor that *coexist* in the two-phase region. If these densities are plotted as a function of temperature the curves AC and BC in Figure 22 are obtained. Points A and B represent the densities

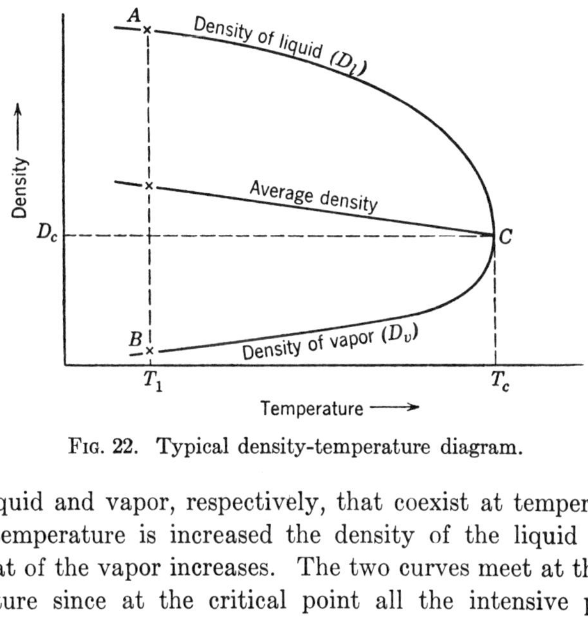

FIG. 22. Typical density-temperature diagram.

of the liquid and vapor, respectively, that coexist at temperature T_1. As the temperature is increased the density of the liquid decreases while that of the vapor increases. The two curves meet at the critical temperature since at the critical point all the intensive properties (density included) of the liquid and vapor are identical.

It has been found that, if the *average* density of the liquid and vapor are plotted as a function of temperature, a straight line results. This is known as the Law of Rectilinear Diameters and may be stated mathematically as

$$\frac{D_l + D_v}{2} = aT + b$$

where D_l and D_v are the densities of the liquid and vapor, respectively, and a and b are two constants that determine the slope and intercept of the straight line obtained by plotting the average density as a function of temperature.

This diagram is useful in calculating the critical volume from density data. The experimental determination of the critical volume is sometimes difficult since it requires the precise measurement of a volume at a high temperature and pressure. However, it is apparent that the straight line obtained by plotting the average density versus temperature intersects the critical temperature at the critical density. The molal critical volume is obtained by dividing the molecular weight by the critical density, or

$$V_c = \frac{\text{MW}}{D_c}$$

This density-temperature diagram can also be used to determine the state of a single-component system. Suppose the *overall* density of the system is known at a given temperature. If this overall density is less than D_v it is obvious that the system is composed entirely of vapor. Similarly, if the overall density is greater than D_l the system is composed entirely of liquid. If, however, the overall density is between D_l and D_v it is apparent that both liquid and vapor are present. In order to calculate the weights of liquid and vapor present one can set up the following volume and weight balances.

$$\text{Volume of liquid} + \text{volume of vapor} = \text{total volume of system}$$

$$\text{Weight of liquid} + \text{weight of vapor} = \text{total weight of system}$$

Let Wt = total weight of a system of known total volume and Wt_v = weight of vapor. Then $\text{Wt} - \text{Wt}_v$ = weight of liquid. Consequently the volume of the vapor is Wt_v/D_v and the volume of the liquid is

$$\frac{\text{Wt} - \text{Wt}_v}{D_l}$$

Substitution in the above volume balance gives

$$\frac{\text{Wt} - \text{Wt}_v}{D_l} + \frac{\text{Wt}_v}{D_v} = \text{total volume} \quad (1)$$

This equation can be solved for Wt_v, provided Wt, D_l, D_v, and the total volume of the system are known.

Example. Ten pounds of a hydrocarbon are placed in a one cubic foot vessel at 60° F. The densities of the coexisting liquid and vapor are known to be 25 lb/cu ft and 0.05 lb/cu ft, respectively, at this temperature. Calculate the weights and volumes of the liquid and vapor phases.

Since the overall density is 10 lb/cu ft the system must be made up of both liquid and vapor since this value is between D_l and D_v. Substitution in equation 1 gives

$$\frac{10 - Wt_v}{25} + \frac{Wt_v}{0.05} = 1 \qquad \text{or} \qquad Wt_v = 0.030 \text{ lb}$$

Therefore, this system consists of 0.030 lb of vapor and 9.97 lb of liquid. The volume of vapor present is 0.030/0.05 = 0.60 cu ft. The volume of the liquid is 0.40 cu ft.

TWO-COMPONENT SYSTEMS

The Pressure-Volume Diagram for a Two-Component System. Consider the pressure-volume diagram of a binary hydrocarbon mixture

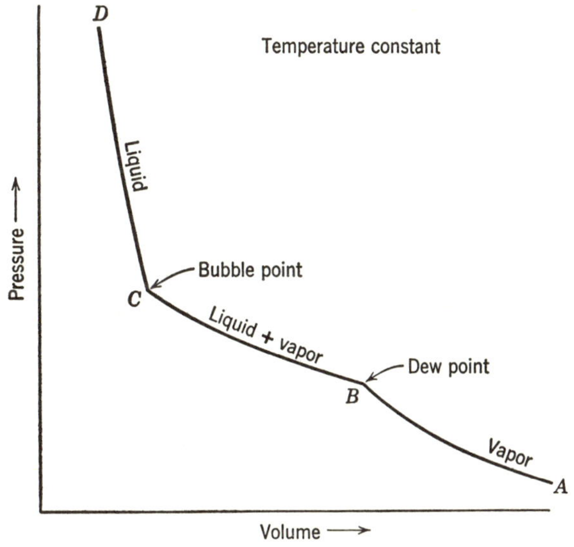

FIG. 23. Pressure volume isotherm for a two-component system.

with a given overall composition. In the following discussion the two components of this mixture will be designated as the more volatile component and the less volatile component, depending on their relative vapor pressures at a given temperature. In Figure 23, the vapor phase isotherm AB and the liquid phase isotherm CD are very similar to those which are obtained for a single-component system. However, the isotherm through the two-phase region is fundamentally different from the corresponding isotherm for a pure component in that the pressure increases as the system passes from the dew point to the bubble point. This is because the compositions of the liquid and the vapor change continuously as the system passes through the two-phase region. At the dew point the composition of the vapor is equal to the overall composition of the system but the infinitesimal amount of liquid which condenses is richer in the less volatile component. However, as more and more liquid is condensed its composition with respect to the *more* volatile component steadily increases (with a corresponding increase in vapor pressure) until the composition of

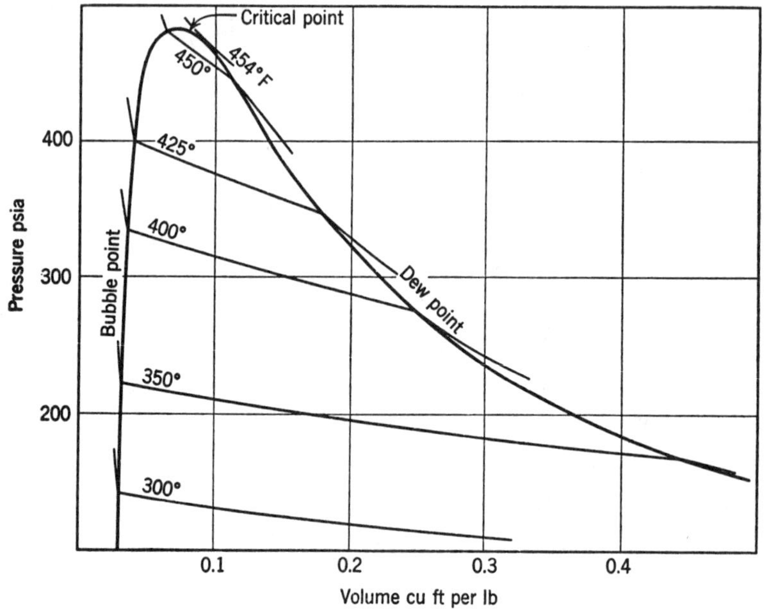

FIG. 24. Pressure-volume diagram for the n-pentane and n-heptane system containing 52.4 weight per cent n-heptane. (Sage and Lacey, *Volumetric and Phase Behavior of Hydrocarbons,* p. 78.)

the liquid becomes equal to that of the system as a whole at the bubble point. The infinitesimal amount of vapor remaining at the bubble point is richer in the *more* volatile component than the system as a whole.

There will be an isotherm similar to *ABCD* for each temperature. The complete P-V diagram for the n-pentane, n-heptane system containing 52.4 weight per cent n-heptane is shown in Figure 24. The critical point is the point where the bubble-point line and dew-point line meet. This is equivalent to the statement that the intensive properties of the coexisting liquid and vapor phases are identical at the critical point. Consequently, the liquid and the vapor are indistinguishable at the critical pressure and temperature. The critical point is no longer at the apex or peak of the two-phase region. This is due to the fact that the isotherms through the two-phase region are not horizontal but have a definite slope. One consequence of this fact is that vapor can exist at pressures above the critical pressure. Similarly, liquids can exist at temperatures greater than the critical temperature. These concepts and their consequences will be discussed in greater detail in a later section.

The Pressure-Temperature Diagram for a Two-Component System. If the bubble-point pressure and the dew-point pressure for the various

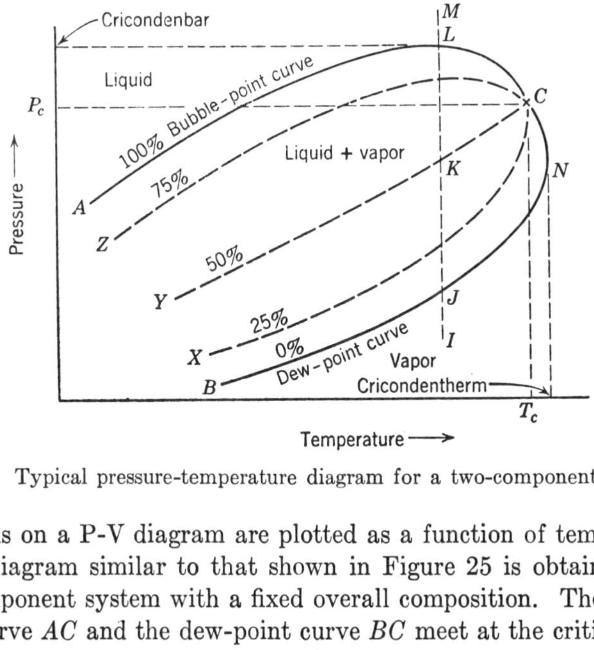

FIG. 25. Typical pressure-temperature diagram for a two-component system.

isotherms on a P-V diagram are plotted as a function of temperature, a P-T diagram similar to that shown in Figure 25 is obtained for a two-component system with a fixed overall composition. The bubble-point curve AC and the dew-point curve BC meet at the critical point C. The critical pressure and temperature are given by P_c and T_c, respectively. Points within the loop ACB represent systems consisting of two phases. Points below the dew-point curve represent vapor and points above the bubble-point curve represent liquid. As in single-component systems points far removed from the two-phase region represent fluid. The P-T diagram shown in Figure 25 is to be contrasted with that of a single-component system in the liquid-vapor region. For the latter, it will be recalled, the bubble-point curve and dew-point curve coincide.

The P-T diagram indicates the phase changes that occur when the pressure and temperature of a system are varied. For example, if a system originally at point I is compressed isothermally at a temperature below T_c along the path IM the following phase changes occur. The system is originally in the vapor state. At the dew point J liquid begins to form and in passing from J to L more and more liquid condenses. At the bubble point L the system is essentially all liquid and only an infinitesimal amount of vapor remains. At the point M the system is in the liquid state.

Sometimes the liquid-vapor volume distribution in the two-phase region is also indicated on P-T diagrams. This can be accomplished by a series of curves each of which represents a certain percentage by volume of liquid. Thus the dotted curves XC, YC, and ZC represent 25%, 50%, and 75% by volume of liquid, respectively. In the isothermal compression described above, the point K would represent 50% liquid and 50% vapor by volume. Obviously, the dew-point curve and the bubble-point curve represent 0% and 100% liquid, respectively.

The P-T diagram clearly shows that liquid can exist above the critical temperature. The highest temperature at which liquid can exist is known as the cricondentherm and is given by the temperature which is tangent to the two-phase loop at N. Similarly, vapor can exist above the critical pressure. The maximum pressure at which vapor can exist is known as the cricondenbar.

Retrograde Phenomena. Consider the behavior of a two-component system in the vicinity of the critical point in greater detail. On the P-T diagram shown in Figure 26 the bubble-point and dew-point curves meet at the critical point C. Consider an isothermal compression along the path AE. The point A which is above the critical temperature, but below the cricondentherm, represents a system in the vapor phase. At the dew point B liquid will begin to condense. More and more liquid will separate as the pressure is increased. However, at point E the dew-point line must be crossed again. This means that all the liquid which formed must *vaporize* since the system is essentially all vapor at the dew point. Consequently, at some point between B and E, at point D for example, the amount of liquid is at a maximum. Therefore in going from D to E liquid vaporizes as the pressure is increased. Since this is the reverse of the behavior at temperatures less than the critical this process is described as isothermal *retrograde* vaporization. The reverse process in going from E to D is known as isothermal retrograde condensation since it involves the formation of liquid with an isothermal decrease in pressure.

Similar phenomena occur at pressures greater than P_c, but less than the cricondenbar. Consider an isobaric temperature increase along the path JG. At the bubble point I the liquid will begin to vaporize but on crossing the bubble point again at G the vapor formed must condense. If H represents the point where the amount of vapor is a maximum the path from H to G represents isobaric retrograde condensation since vapor condenses as the temperature is increased. The reverse process from G to H is known as isobaric retrograde vaporization.

In other words, retrograde condensation is defined as the formation of liquid by an isothermal decrease in pressure or an isobaric increase

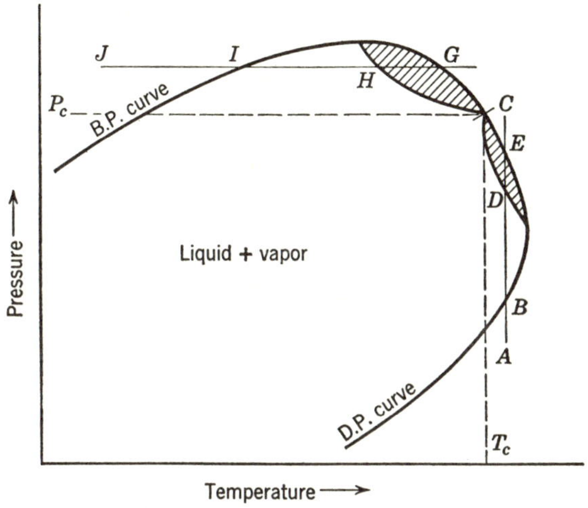

FIG. 26. Pressure-temperature diagram in the vicinity of the critical point of a two-component system which exhibits retrograde phenomena.

in temperature. Similarly, retrograde vaporization is the formation of vapor by an isothermal compression or an isobaric decrease in temperature. Retrograde phenomena can only occur in the shaded areas in Figure 26. Obviously they occur only at pressures between P_c and the cricondenbar or at temperatures between T_c and the cricondentherm. One may think of these phenomena as being abnormal but in reality this behavior is characteristic of almost all systems composed of two or more components. Indeed, this behavior was predicted by early investigators before it was observed experimentally. However, this behavior need not be general since retrograde phenomena would not occur in a system whose dew-point curve and bubble-point curve meet in an acute angle at the critical point so that the cricondentherm and the cricondenbar are equal to T_c and P_c respectively.

Figure 27 shows the retrograde regions on a P-V diagram. This diagram shows typical isotherms in the region of the critical point C. T_c is the isotherm at the critical temperature and T_1 represents the cricondentherm. Beginning at point A the isotherm at temperature T, which is above T_c but below T_1, passes through the two-phase region along the path ABD. At the dew point A liquid begins to form but at the second dew point D only an infinitesimal amount of liquid re-

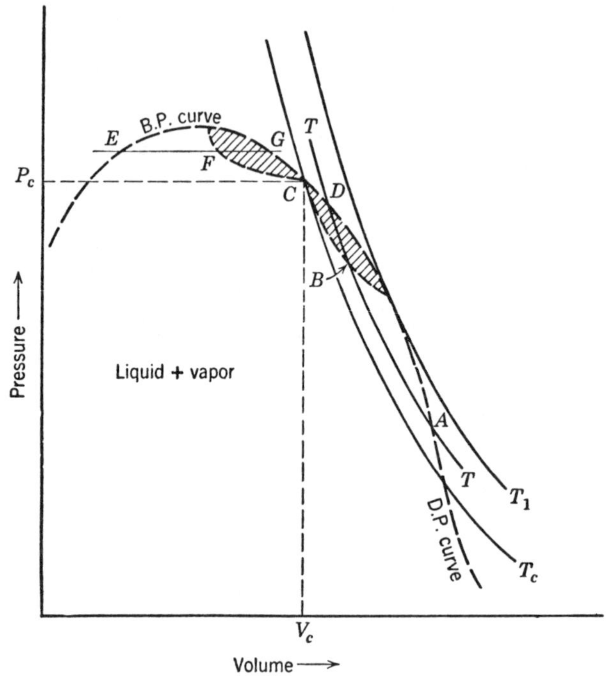

FIG. 27. Typical pressure-volume diagram for a two-component system in the region of the critical.

mains. If the maximum amount of liquid occurs at B then isothermal retrograde vaporization occurs from B to D. Similarly, for an isobaric increase in temperature along the path EFG, retrograde condensation occurs from F to G if F represents the point where the amount of vapor is a maximum. The shaded areas in the diagram represent the areas where retrograde phenomena occur.

Unfortunately, the terminology applied to retrograde phenomena has not yet been standardized in the literature. The phenomena described as retrograde vaporization in the preceding pages are sometimes referred to as retrograde condensation and vice versa. Furthermore, the entire two-phase region between the cricondentherm and the critical temperature and between the cricondenbar and the critical pressure is sometimes referred to as the retrograde region. The definitions given in this text are in accord with the most common usage in petroleum engineering practice.

The Composite P-T Diagram. The P-T diagram previously described represents the phase behavior of a two-component system with

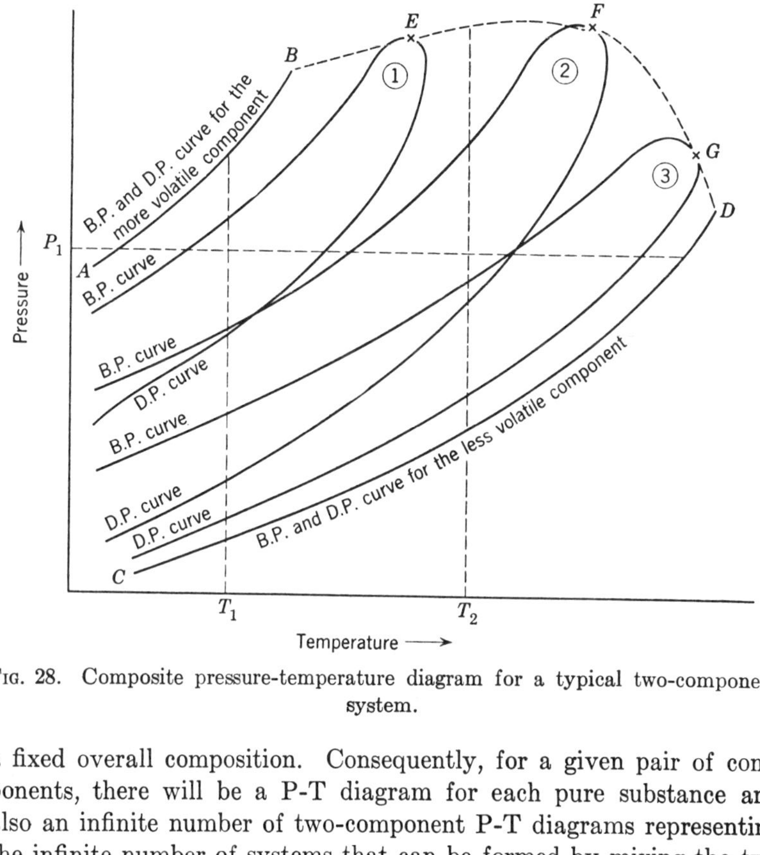

FIG. 28. Composite pressure-temperature diagram for a typical two-component system.

a fixed overall composition. Consequently, for a given pair of components, there will be a P-T diagram for each pure substance and also an infinite number of two-component P-T diagrams representing the infinite number of systems that can be formed by mixing the two components in various ratios. If the bubble-point and dew-point curves for the two pure components and for several mixtures are plotted on the same P-T diagram as in Figure 28 the composite P-T diagram is obtained. The curves *AB* and *CD* represent the vapor pressures as a function of temperature of the pure more volatile component and the pure less volatile component respectively. The three loops shown in this diagram, numbered 1, 2, and 3, define the two-phase regions for three of the infinite number of mixtures which can be formed with the two pure components. Loop 1 represents the

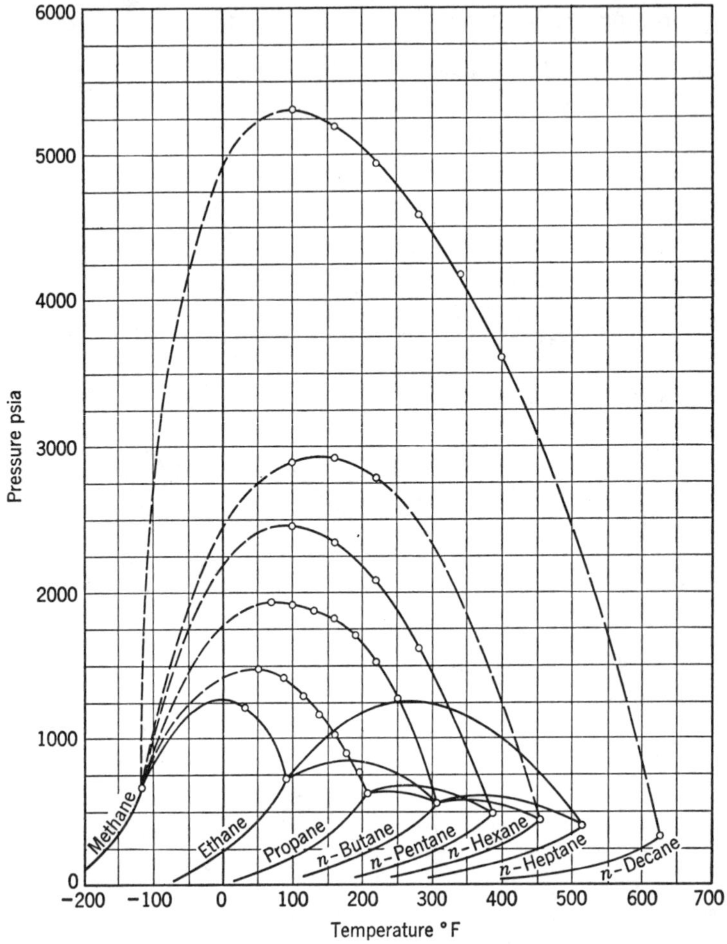

Fig. 29. Critical loci of binary *n*-paraffin systems. (Brown, Katz, Oberfell, and Alden, *Natural Gas and Volatile Hydrocarbons,* 1948, p. 4.)

two-phase region for a system which is relatively rich in the more volatile component. Similarly, loop 3 represents a system which is rich in the less volatile component. Finally loop 2 represents a system in which the concentrations of the two components are more nearly equal. This diagram describes the phase behavior of two-component mixtures as a function of composition. The dotted line through the critical points B, E, F, G, and D is known as the critical locus and represents the position of the critical point as a function of composition of the system.

The critical loci of binary systems composed of normal paraffin hydrocarbons are shown in Figure 29. This diagram clearly shows that the critical pressure for a mixture can be greater than the critical pressure of each pure component. It will be recalled that the pseudo-critical pressure used in correcting for the non-ideal behavior of gases (equation 20, Chapter 2) is by definition *between* the critical pressures of the constituents in a two-component system. Obviously, there is no relation between the actual critical pressure of a mixture and the pseudo-critical pressure. Similarly, there is no relation between the actual critical temperature and pseudo-critical temperature.

The Pressure-Composition Diagram for a Two-Component System. Another way of describing the phase behavior of a binary system as a function of composition is by means of a pressure-composition diagram. This diagram describes the behavior of a two-component system at a fixed temperature. It is constructed by plotting the dew-point pressure and bubble-point pressure as a function of composition. The bubble-point line is drawn through the points which represent the bubble-point pressures as a function of composition. Similarly, the dew-point line is drawn through the points which represent the dew-point pressures. Figure 30 represents a typical pressure-composition diagram for a hydrocarbon system. The points A and B represent the vapor pressures, the dew-point pressures, and the bubble-point pressures for the pure more volatile component and the pure less volatile component, respectively. Similarly, the points C and D represent the dew-point pressure and the bubble-point pressure for a mixture which contains 25% by weight of the less volatile component and 75% by weight of the more volatile component. The area below the dew-point line represents vapor, the area above the bubble-point line represents liquid, and the area between these two curves represents the two-phase region, where liquid and vapor coexist.

The pressure-composition diagram is related to the composite pressure-temperature diagram previously described in the following manner. In Figure 28 let T_1 represent the fixed temperature at which the pressure-composition diagram is to be constructed. This isotherm intersects the two-phase loop of a particular composition at the bubble-point pressure and the dew-point pressure. If these two pressures and the composition are plotted, two points on the pressure-composition diagram are obtained. Other pairs of points can be obtained in the same way at other compositions but at the same constant temperature.

In the diagram shown in Figure 30 the composition is expressed in weight per cent of the less volatile component. It is to be understood that the composition may equally well be expressed in terms of weight per cent of the more volatile component in which case the bubble-point and dew-point lines have the opposite slope. Furthermore, the

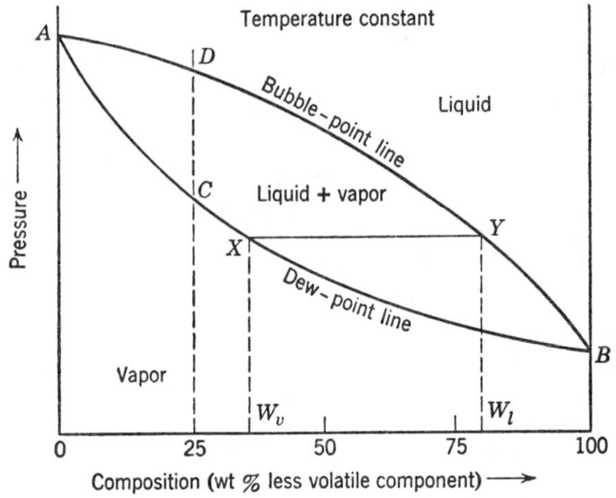

FIG. 30. Typical pressure-composition diagram for a two-component system. Composition expressed in terms of Wt % of the less volatile component.

composition may be expressed in terms of mole per cent or mole fraction if desired.

Consider the meaning of the horizontal line XY through the two-phase region in this diagram. The points X and Y at the extremities of this line represent a vapor and a liquid that exist at the same temperature (temperature is constant in this diagram) and the same pressure (XY is horizontal). *Consequently, the points* X *and* Y *represent the compositions of coexisting liquid and vapor phases.* In other words, at a constant temperature, and at the pressure represented by the horizontal line XY, the compositions of the vapor and liquid that coexist in the two-phase region are given by W_v and W_l, respectively. W_v and W_l represent the weight percentages of the less volatile component in the vapor and liquid, respectively.

In order to more fully understand the meaning of a pressure-composition diagram the behavior of a system which is initially in the vapor phase and which is subjected to an isothermal compression through the two-phase region will be described. In the pressure-composition diagram shown in Figure 31 the composition is expressed in *mole fraction* of the *more* volatile component. If the pressure is increased on a system represented by point A, no phase change occurs until the dew point B is reached at pressure P_1. At the dew point an infinitesimal amount of liquid forms whose composition is given by x_1. The composition of the vapor is still equal to the original composition z. As the pressure is increased more liquid forms and the

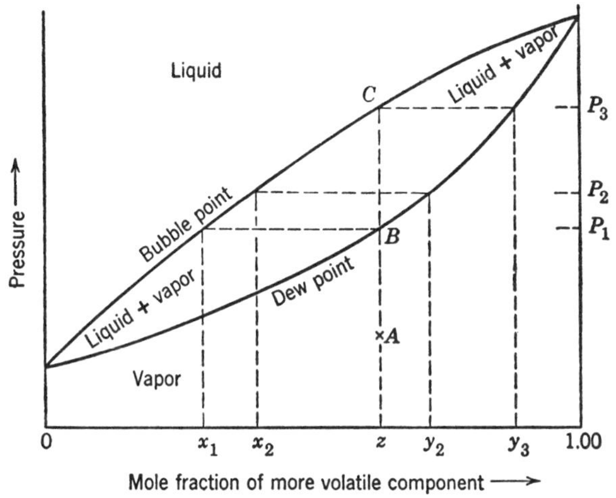

FIG. 31. Pressure-composition diagram illustrating an isothermal compression through the two-phase region.

compositions of the coexisting liquid and vapor are given by projecting the ends of the straight, horizontal line through the two-phase region of the composition axis. For example, at P_2 both liquid and vapor are present and the compositions are given by x_2 and y_2. At pressure P_3 the bubble point C is reached. The composition of the liquid is equal to the original composition z. The infinitesimal amount of vapor still present at the bubble point has a composition given by y_3.

As previously indicated the extremities of a horizontal line through the two-phase region represent the compositions of coexisting phases. However, the *composition* of a phase and the *amount* of a phase present in a two-phase system should not be confused. At the dew point, for example, there is only an infinitesimal amount of liquid present but it consists of finite mole fractions of the two components. An equation for the relative amounts of liquid and vapor in a two-phase system may be derived as follows.

Let n = total number of moles in system
n_l = moles of liquid
n_v = moles of vapor
z = mole fraction of more volatile component in the system
x = mole fraction of more volatile component in the liquid
y = mole fraction of more volatile component in the vapor

Then nz = moles of more volatile component in the system
$n_l x$ = moles of more volatile component in the liquid
$n_v y$ = moles of more volatile component in the vapor

Since the number of moles of the more volatile component in the liquid and in the vapor are equal to the number of moles of the more volatile component in the system, one has

$$nz = n_l x + n_v y \tag{2}$$

Furthermore it is obvious that

$$n_l = n - n_v$$

Substituting in equation 2

$$nz = (n - n_v)x + n_v y$$

and rearranging one obtains

$$\frac{n_v}{n} = \frac{z - x}{y - x} \tag{3}$$

Similarly, if n_v is eliminated in equation 2 instead of n_l one obtains

$$\frac{n_l}{n} = \frac{z - y}{x - y} \tag{4}$$

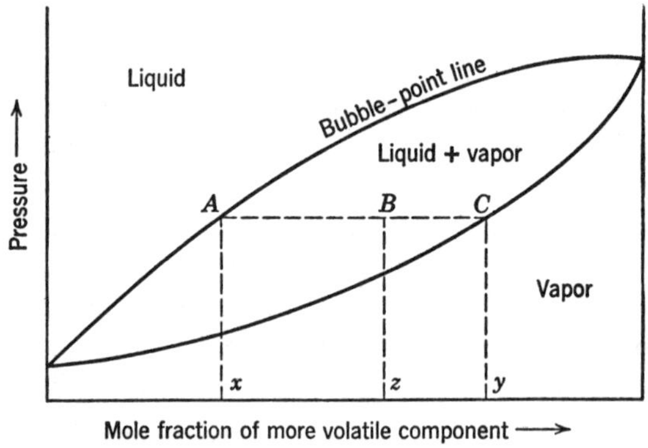

Fig. 32. Geometrical interpretation of equations for the amount of liquid and vapor in the two-phase region.

The geometrical interpretation of equations 3 and 4 is shown in Figure 32. Since $z - x = AB$ and $y - x = AC$ equation 3 becomes

$$\frac{n_v}{n} = \frac{z - x}{y - x} = \frac{AB}{AC}$$

Similarly, equation 4 becomes

$$\frac{n_l}{n} = \frac{BC}{AC}$$

That is, the number of *moles* of vapor is to the total number of moles in the system as the length of the line segment AB is to the length AC. Also the number of *moles* of liquid is to the total number of moles in the system as the length of the line segment BC is to the length AC. A convenient way to remember these equations is that in calculating the amount of a phase, use the length of the line segment furthest from this phase on the phase diagram over the total length of the line in the two-phase region. The reader should convince himself that the results would have been the same if the mole fraction of the less volatile component had been plotted on the phase diagram instead of mole fraction of the more volatile component.

EXAMPLE. A system is composed of three moles of isobutane and one mole of normal heptane. The system is separated at a fixed temperature and pressure and a liquid and vapor phase recovered. The mole fraction of isobutane in the recovered liquid is 0.370 and 0.965 in the recovered vapor. Calculate the quantity of liquid and vapor recovered on a molal basis.

Since $z = 0.750$, $x = 0.370$, $y = 0.965$, and $n = 4$ substitution in equation 3 gives

$$n_v = n\left(\frac{z - x}{y - x}\right) = 4\left(\frac{0.750 - 0.370}{0.965 - 0.370}\right) = 2.56 \text{ moles of vapor}$$

The quantity of liquid is

$$n_l = n - n_v = 4 - 2.56 = 1.44 \text{ moles of liquid}$$

n_l could also have been obtained by substitution in equation 4.

If the composition is expressed in weight fraction instead of mole fraction the following equations may be derived for the weights of liquid and vapor.

Let Wt = total weight of system
Wt_l = weight of liquid
Wt_v = weight of vapor

and

w_o = weight fraction of more volatile component in original system
w_l = weight fraction of more volatile component in liquid
w_v = weight fraction of more volatile component in vapor

A material balance on the more volatile component leads to the following equations

$$\frac{\mathrm{Wt}_v}{\mathrm{Wt}} = \frac{w_o - w_l}{w_v - w_l} \quad \text{and} \quad \frac{\mathrm{Wt}_l}{\mathrm{Wt}} = \frac{w_o - w_v}{w_l - w_v} \tag{5}$$

Furthermore, the geometrical interpretation of these equations is the same as before. However, it should be emphasized that if the composition is expressed in weight fraction or weight per cent the ratios $\mathrm{Wt}_l/\mathrm{Wt}$ and $\mathrm{Wt}_v/\mathrm{Wt}$ represent *weight* of liquid over total *weight* and *weight* of vapor over total *weight*, respectively. If, on the other hand, composition is expressed in mole fraction or mole per cent, then, as shown before, n_l/n and n_v/n represent *moles* of liquid over total *moles* and *moles* of vapor over total *moles* respectively.

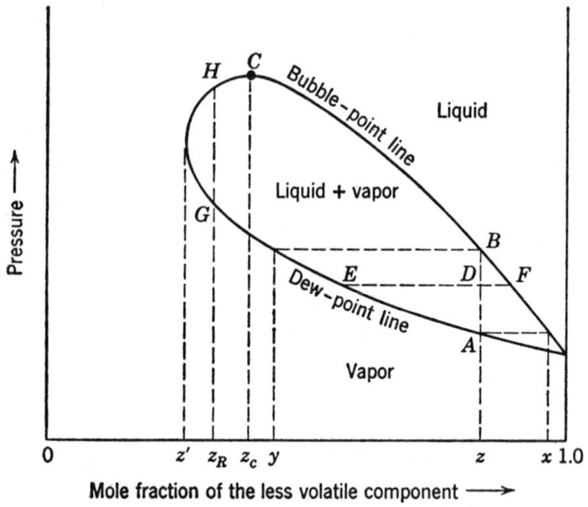

FIG. 33. Typical pressure-composition diagram for a two-component system at a temperature between the critical temperatures of the two pure components.

The pressure-composition diagram shown in Figure 30 was constructed at a temperature less than the critical temperature of each pure component. If the temperature is fixed at a value between the criticals for the components, such as T_2 in Figure 28, then a pressure-composition diagram similar to that shown in Figure 33 is obtained. The bubble-point line and the dew-point line meet at point C which represents the critical pressure for the system whose composition is given by z_c. No two-phase region exists for systems with a mole fraction of the less volatile component below z'. The composition z' is that of the system whose cricondentherm is equal to the temperature T_2 at which the pressure-composition diagram was constructed.

The interpretation of this diagram is similar to that of the pressure-composition diagrams previously described. Thus, for the system whose initial overall composition is z the dew point is at A and the bubble point is at B. The composition of the liquid at the dew point is x and the composition of the vapor at the bubble point is y. At point D between the dew point and the bubble point two phases exist and the ratio between the moles of liquid and the moles of vapor is given by ED/DF. A system whose initial composition is between z' and z_c exhibits retrograde phenomena when subjected to an isothermal compression. For example, a system whose overall composition is given by z_R is at the dew point at G and again at H. As this system is compressed isothermally and passes from G to H the amount of liquid increases until it reaches a maximum and then decreases to zero again. The vaporization of liquid which occurs as the pressure is increased and the point H is approached will be recognized as isothermal retrograde vaporization.

The Temperature-Composition Diagram for a Two-Component System. Consider again the composite P-T diagram shown in Figure 28.

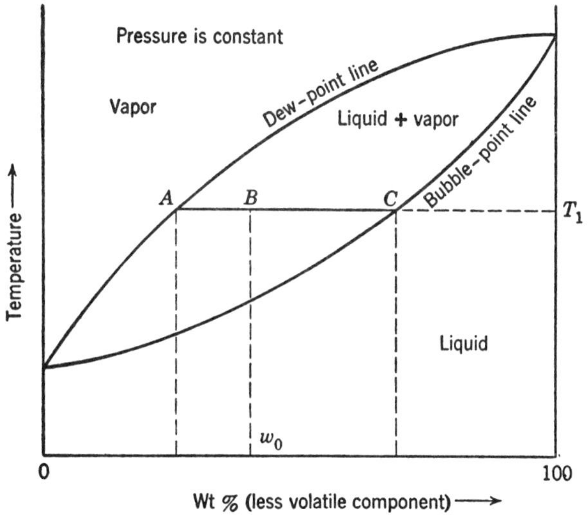

FIG. 34. Illustrative temperature-composition diagram for a binary hydrocarbon system.

At a given pressure, P_1, for example, there will be a dew-point temperature and a bubble-point temperature for each composition. If these temperatures are plotted as a function of composition a temperature-composition similar to that shown in Figure 34 is obtained.

Here again, the composition may be expressed in terms of weight per cent or mole per cent of either the more volatile or the less volatile component. The area above the dew-point temperature curve represents vapor. The area below the bubble-point temperature curve represents liquid. The area between these two curves represents the two-phase region. In this diagram, the extremities of a horizontal line through the two-phase region will again represent the compositions of coexisting phases. If 14.7 psia is chosen as the fixed pressure, the bubble-point temperature line will represent the normal boiling-point line since the normal boiling point is defined as the temperature at which the bubble-point pressure is one atmosphere.

It can be shown that the amounts of liquid and vapor that coexist in the two-phase region can be represented by the lengths of the horizontal line segments through the two-phase region just as before. Thus, if the overall composition is represented by weight per cent w_o and the temperature is T_1 then the ratio of weight of liquid to the total weight of the system is AB/AC. Similarly, the ratio of weight of vapor to the total weight is BC/AC.

MULTICOMPONENT SYSTEMS

The phase behavior of multicomponent hydrocarbon systems in the liquid-vapor region is very similar to that of binary systems. However, it is obvious that two-dimensional pressure-composition and temperature-composition diagrams no longer suffice to describe the behavior of multicomponent systems. For a multicomponent system with a given overall composition, the characteristics of the P-T and P-V diagrams are very similar to those of a two-component system. For systems involving crude oils which usually contain appreciable amounts of relatively non-volatile constituents, the dew points may occur at such low pressures that they are practically unattainable. This fact will modify the behavior of these systems to some extent.

The P-V Diagram for a Multicomponent System. For a relatively volatile multicomponent system, a gasoline for example, an isotherm on the P-V diagram is similar to its counterpart for a binary system (Figure 23). However, it is commonly found that at the dew point the break in the P-V isotherm is not very pronounced in multicomponent systems. Consequently, for systems of this type, it may be very difficult to fix the dew point in this manner. This experimental difficulty can be overcome by using a windowed cell and observing the pressure and volume when traces of liquid appear in the system.

A typical P-V isotherm for a crude-oil system is shown in Figure 35. The point *A* represents an entirely liquid system at a relatively high pressure. As the pressure is decreased isothermally the bubble point is reached at *B*. The bubble-point pressure is usually designated as the saturation pressure (P_s) for crude-oil systems. This is due to the fact that in crude oils it is customary to regard the vapor phase which forms at the bubble point as having been gas dissolved in the liquid phase. This is entirely justifiable and the formation of vapor in *any* system can be regarded as vapor coming out of solution.

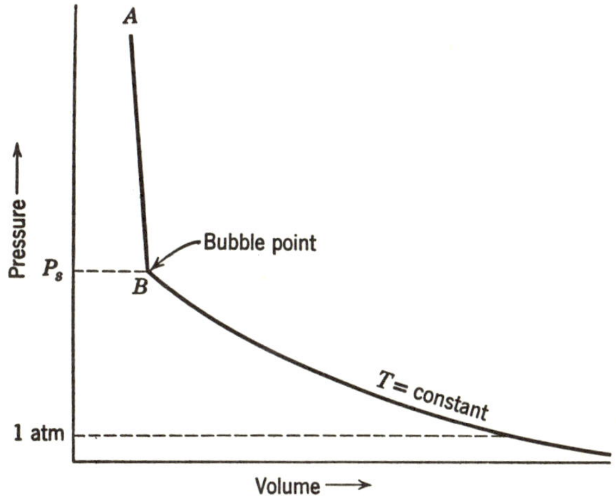

FIG. 35. Pressure-volume isotherm for a crude oil.

Consequently, at the saturation pressure, the liquid is regarded as being saturated with gas and any further decrease in pressure results in the liberation of solution gas, that is, the formation of a vapor phase. As the pressure is decreased below P_s more and more vapor forms (more and more gas comes out of solution). When atmospheric pressure is reached a crude-oil system generally consists of both liquid and vapor. To vaporize the system completely may require an extremely low pressure so that the dew point is practically unattainable.

The P-T Diagram for a Multicomponent System. As previously stated, the characteristics of a P-T diagram for a multicomponent system and for a two-component system are very similar. These multicomponent P-T diagrams are useful in describing the phase behavior of petroleum reservoirs. Consider a hydrocarbon mixture with a given, overall composition whose P-T diagram is shown in Figure 36. This diagram, which indicates extremely low dew-point pressures at low temperatures, may represent a crude-oil system. If the surface temperature and pressure are indicated by point A (P_1, T_1) and the reservoir temperature and pressure by point B (P_0, T_0), this diagram would represent a reservoir with a gas cap containing both liquid and

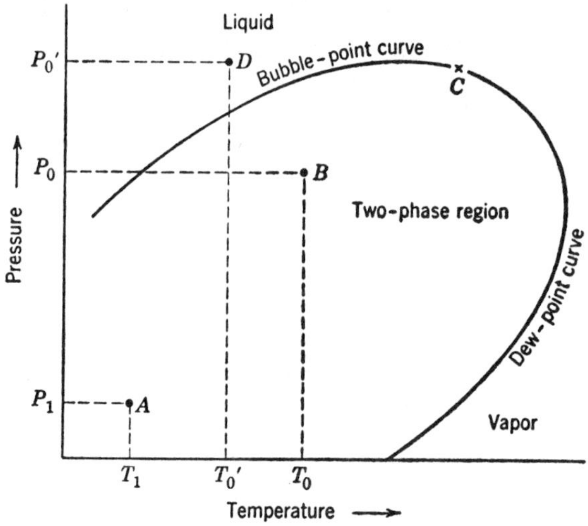

FIG. 36. Pressure-temperature diagram for a multicomponent system whose properties are similar to those of a crude oil.

vapor which produces both liquid and gas. If the reservoir temperature and pressure is represented by point D (P_0', T_0'), this diagram would represent an undersaturated crude (i.e., all liquid with no free gas present) which produces both liquid and gas.

In the P-T diagram shown in Figure 37 the point E represents a dry gas reservoir. If the surface conditions are at F this reservoir

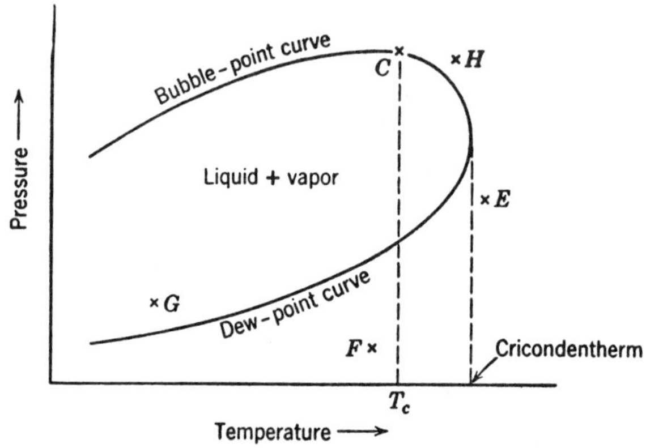

FIG. 37. Multicomponent pressure-temperature diagram illustrating the phase behavior of typical petroleum reservoirs.

would produce dry gas. If, on the other hand, the surface conditions are at G this reservoir would produce both liquid and gas and would be known as a condensate reservoir. A reservoir whose initial temperature and pressure are at H and the surface conditions at F is known as a retrograde condensate reservoir since in producing this reservoir the fluid would pass through the retrograde region. If the pressure gradient is such that retrograde condensation occurs in the reservoir two phases will form there, but a further decrease in pressure will cause the more dense liquid phase to vaporize again.

The Gibbs' Phase Rule. Much of the qualitative information concerning phase behavior described in the preceding sections of this chapter may be summarized in a single generalization known as the Gibbs' phase rule. This generalization was discovered by J. Willard Gibbs in 1876 and deals with the number of phases that can coexist in equilibrium for a system under given conditions of temperature and pressure.

Before stating the phase rule it would be advantageous to reconsider the phase behavior of a single-component system from a slightly different standpoint. If a single-component system exists as a single phase it is necessary to specify both the temperature and the pressure in order to define the system insofar as its *intensive* properties are concerned. Such a system is said to be bivariant or to have two degrees of freedom. However, if a single-component system is in the two-phase region it is necessary to specify only one variable to define it. Thus, when both liquid and vapor are present at a specified temperature the pressure is fixed at the vapor pressure and all the intensive properties of both the liquid and vapor phases are defined. Conversely, if the pressure is specified the temperature is also defined as the temperature at which vapor pressure is equal to the specified pressure. Consequently, a two-phase, single-component system is univariant or is said to have one degree of freedom. Finally, if three phases are present in a single-component system then the system is at the triple point and is said to be invariant since the intensive properties of all three phases are defined.

Gibbs' phase rule may be stated in equation form as

$$F = C - P + 2 \tag{6}$$

where F is the variance or the number of degrees of freedom, C is the minimum number of components or chemical compounds required to make up the system, and P is the number of phases that are present when the system is at equilibrium. This equation, which is valid for systems whose phase behavior is determined by temperature, pressure, and composition, is presented without derivation. (For derivation, see, for example, MacDougall, *Thermodynamics and Chemistry.*) This equation states that a single-component system ($C = 1$) is bivariant ($F = 2$) when one phase is present ($P = 1$), univariant ($F = 1$) when two phases are present ($P = 2$), and invariant ($F = 0$) when three phases are present ($P = 3$). Consequently, it is apparent that the phase rule predicts the phase behavior of a system insofar as its variance is concerned.

It should be emphasized that the phase rule is only qualitative and does not predict the actual values of the intensive properties of the system. Neither does it predict the relative amounts of the phases present in a system.

The phase rule as defined by equation 6 is applicable to systems containing any number of components. For a hydrocarbon mixture in which there is no chemical interaction between the constituents, the value of C in equation 6 is simply equal to the number of components in the mixture. Consider a two-component system in the two-phase region. According to the phase rule this system is bivariant. This is in agreement with the conclusions reached previously since it has been shown that it is necessary to specify both the temperature and the pressure in order to define the system intensively. If these two variables are fixed the system will consist of a liquid and a vapor phase whose compositions and intensive properties are defined. (See Figure 30.) If, on the other hand, a two-component system exists in a single-phase region its variance is three according to the phase rule. It has been shown that all three variables (i.e., temperature, pressure, and overall composition) must be specified in order to define the intensive properties of the system in this case.

The phase rule is a useful tool which can be used to correlate and summarize the phase behavior of systems composed of one or more phases in equilibrium with one another. In this respect it is particularly useful in predicting the qualitative behavior of multicomponent, multiphase systems whose phase behavior can be extremely complex.

REFERENCES

Calhoun, J., *Fundamentals of Reservoir Engineering,* University of Oklahoma Press, Norman, Okla. (1953).

Campbell, A. N., and N. O. Smith, *The Phase Rule and Its Applications,* Dover Publications, Inc., New York (1951).

Daniels, F., *Outlines of Physical Chemistry,* John Wiley & Sons, New York (1948).

MacDougall, F. H., *Thermodynamics and Chemistry,* John Wiley & Sons, New York (1939).
Muskat, M., *Physical Principles of Oil Production,* McGraw-Hill Book Co., New York (1950).
Pirson, S., *Elements of Oil Reservoir Engineering,* McGraw-Hill Book Co., New York (1950).
Sage, B., and W. Lacey, *Volumetric and Phase Behavior of Hydrocarbons,* Stanford University Press, Stanford University, Calif. (1939).

PROBLEMS

1. The densities of the coexisting vapor and liquid of a pure compound at various temperatures are as follows:

$t°$ C	30	50	70	100	120
D_l (grams/cc)	0.6455	0.6116	0.5735	0.4950	0.4040
D_v (grams/cc)	0.0142	0.0241	0.0385	0.0810	0.1465

If the critical temperature is 126.9° C, what is the molal critical volume if the molecular weight is 50? If 300 grams are placed in a 1-liter vessel at 30° C, calculate the weights of liquid and vapor present. Calculate the same quantities if only 10 grams are placed in the vessel. *Answer:* $V_c = 186$ cc.

2. A vessel of 50 cu ft capacity is evacuated and thermostated at 60° F. Five pounds of liquid propane are injected. What will be the pressure in the vessel and what will be proportions of liquid and vapor present? Repeat calculations for 100 lb of propane injected. The densities of coexisting liquid and vapor propane at 60° F are 31.75 lb/cu ft and 0.990 lb/cu ft, respectively.

3. A tank of butane (containing both liquid and vapor at 60° F) has a gage pressure of 15 psig. What percentage of the total volume of vapor is occupied by an inert gas, if atmospheric pressure is 14.5 psia? If the inert gas is nitrogen, what are the weight fractions in the vapor?

4. A cylinder contains air at a pressure of 20 psia at 60° F. A second cylinder of double capacity contains pure pentane with a *small* amount of liquid pentane present. If these two tanks are interconnected, what is the final pressure under the above conditions of temperature and pressure? What are mole, volume, and weight fractions?

Note: A *small* amount of liquid means that its volume can be neglected but sufficient is present to saturate both tanks with vapor. *Answer:* $P = 13.7$ psia.

5. Derive equations for the amounts of liquid and vapor present in the two-phase region when the composition is expressed in terms of (1) weight fraction of the more volatile component, (2) weight fraction of the less volatile component.

6. A gas consists of a 50–50 mixture by weight of two hydrocarbons. The pressure is increased isothermally until two phases appear. The liquid phase consists of 40% by weight of the more volatile constituent and the vapor phase contains 65% by weight of the more volatile constituent. If the total weight is 80 lb, what are the weights of the liquid and vapor phases? *Answer:* $\text{Wt}_l = 48$ lb, $\text{Wt}_v = 32$ lb.

7. Describe the changes in composition and amount of liquid and vapor that occur on an isobaric temperature increase through the two-phase region of a temperature-composition diagram.

8. A system composed of ethane hydrate, water, and ethane is classed as a two-component system when Gibbs' phase rule is applied since it could be formed from water and ethane. What is the variance of this system when a solid, a liquid, and a vapor phase coexist in equilibrium? If the temperature of this three-phase system is specified, would it be possible to alter the pressure without the disappearance of a phase?

CHAPTER 5

QUANTITATIVE PHASE BEHAVIOR

The qualitative phase behavior of hydrocarbon systems was described in the previous chapter. The quantitative treatment of these systems will now be discussed and the methods for calculating their phase behavior presented. It will become apparent that the liquid and vapor phases of mixtures of two or more hydrocarbons are in reality solutions (see below), so that it will be necessary to discuss the laws of solution behavior. Analogous to the treatment of gases, the behavior of a hypothetical fluid known as a perfect, or ideal, solution will be described. This will be followed by a description of actual solutions and the deviations from ideal solution behavior that occur.

IDEAL SOLUTIONS

Solutions. A solution is defined as a homogeneous mixture of two or more substances, which has the same chemical composition and the same physical properties throughout. All gas mixtures are examples of solutions since gases are completely miscible with one another. Similarly, liquid mixtures of alcohol and water are solutions since they too are homogeneous, single-phase systems. On the other hand, a liquid hydrocarbon and water do not form solutions since these two liquids do not dissolve in one another and a heterogeneous, two-phase system results. In general, the more closely two substances resemble one another chemically, the more likely are they to form a solution.

Because of their chemical similarity, mixtures of hydrocarbons always form solutions that are miscible in all proportions.

When a solution is composed of a small amount of one component and a large amount of a second component it is customary to refer to the former as the solute and the latter as the solvent. However, this distinction is purely arbitrary, particularly when the two components are completely miscible and are present in nearly equal amounts.

Mole per cent or mole fraction, weight per cent or weight fraction, and volume per cent or volume fraction may be employed to designate the composition of a solution. Avogadro's Law is not applicable to liquids, and equal volumes of different liquids do not contain the same number of molecules. Consequently, mole per cent and volume per cent are not equivalent in liquid solutions as they were in perfect-gas mixtures. To convert mole per cent to weight per cent the procedure is identical with that previously described for gases. To calculate the volume per cent of a liquid solution from the mole per cent or weight per cent the densities of the pure components must be known.

Ideal Solutions. In an ideal solution there are no special forces of attraction between the constituent molecules. Thus, in an ideal solution composed of molecules of A and B, the force of attraction between a molecule of A and a molecule of B is the same on the average as that between two molecules of A or two of B. Consequently, there is no heating effect when the components of an ideal solution are mixed. Furthermore, the volume is equal to the sum of the volume of its components, that is

$$V = \Sigma V_i$$

where V_i is the volume of the ith component. Other physical properties of an ideal solution can be calculated by averaging the properties of each constituent in the proper manner. For example, the density is given by

$$D_{\text{solution}} = \Sigma \text{ volume fraction}_i \times D_i^0$$

where D_i^0 is the density of the pure ith component.

No solution is ideal but when the components resemble one another closely the resulting solution is likely to approach the behavior of an ideal solution. Consequently many of the hydrocarbon mixtures which are of particular interest to the petroleum engineer can be expected to follow ideal solution behavior more or less closely under ordinary conditions of temperature and relatively low pressures.

The Vapor Pressure of an Ideal Liquid Solution. The vapor pressure of an ideal solution may be calculated using Raoult's Law. This law states that for an ideal solution the partial pressure of a component in the vapor is equal to the product of the mole fraction of that component in the liquid and the vapor pressure of the pure component. In the form of an equation

$$P_A = x_A P_A^0$$

where P_A is the partial pressure of component A in the vapor, x_A is the mole fraction of component A in the liquid solution, and P_A^0 is the vapor pressure of pure A. In an ideal solution Raoult's Law is applicable to each component so that

$$P_i = x_i P_i^0 \tag{1}$$

where the subscript i designates the ith component of the mixture. Consequently, if the vapor pressure of each component is known, its partial pressure in the vapor may be calculated for a solution of given concentration at a given temperature. The total pressure exerted by the vapor is equal to the sum of the partial pressures of its components, that is,

$$P_T = \Sigma x_i P_i^0 \tag{2}$$

where P_T is the total pressure. This *total pressure* is the *vapor pressure* of the solution. Furthermore, the total pressure is also the *bubble-point pressure* since the application of an infinitesimally greater pressure will result in an all-liquid system.

If the infinitesimal amount of vapor which exists at the bubble point is assumed to be a perfect gas, Dalton's Law of partial pressures (Chapter 2, equation 16) is applicable and

$$P_i = y_i P_T \qquad \text{or} \qquad y_i = \frac{P_i}{P_T} \tag{3}$$

where P_i is the partial pressure of the ith component in the vapor, y_i is the mole fraction of the ith component in the vapor, and P_T is the total pressure. Consequently, the bubble-point pressure of a solution may be calculated using equation 2 and the composition of the vapor at the bubble point may then be calculated using equation 3.

Example. At 0° F calculate the bubble-point pressure and the composition of the vapor at the bubble point for a two-component solution having a mole fraction of propane equal to 0.5 and a mole fraction of butane equal to 0.5. Repeat these calculations for a solution whose mole fraction of propane is 0.25 and whose mole fraction of butane is 0.75.

The vapor pressures of pure propane and butane at 0° F are 38.20 psia and 7.30 psia, respectively. (See Figure 14 and Appendix *B* for vapor pressure data.)

For the solution whose mole fractions of propane and butane are each 0.5 the following calculations apply

Component	P_i^0	x_i	$P_i = x_i P_i^0$	$y_i = P_i/P_T$
C_3H_8	38.20	0.50	19.10	0.840
C_4H_{10}	7.30	0.50	3.65	0.160
			$P_T = 22.75$ psia	

The bubble-point pressure for this solution is 22.75 psia at 0° F. The mole fraction of propane in the vapor is 0.840 and the mole fraction of butane in the vapor is 0.160 at the bubble point.

For the solution whose mole fraction of propane is 0.25 and whose mole fraction of butane is 0.75 a similar calculation gives the following results

Component	P_i^0	x_i	$P_i = x_i P_i^0$	$y_i = P_i/P_T$
C_3H_8	38.20	0.25	9.55	0.635
C_4H_{10}	7.30	0.75	5.48	0.365
			$P_T = 15.03$ psia	

The bubble-point pressure is 15.03 psia at 0° F. The mole fraction of propane and butane in the vapor at the bubble point are 0.635 and 0.365, respectively.

The relationship between the quantities calculated in the example presented above and the phase diagrams previously described should be clearly understood. In Figure 38 the pressure-composition diagram for the propane-butane system is shown. In this diagram the composition is expressed in terms of mole fraction of butane. The points

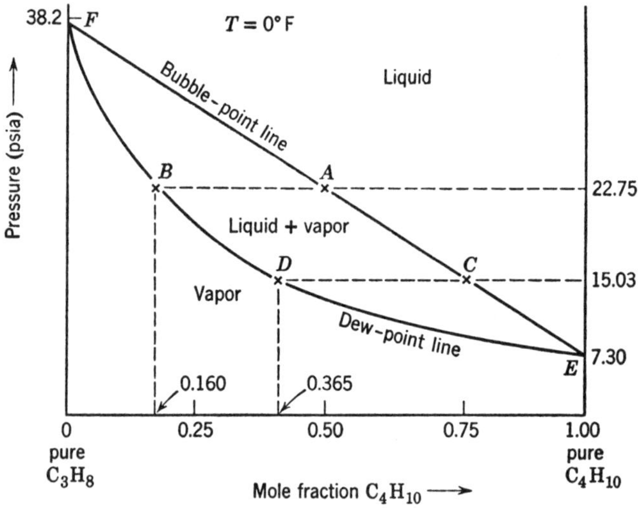

FIG. 38. Pressure-composition diagram for the propane-butane system at 0° F. Calculated assuming ideal solution behavior.

A, B, C, and D have been calculated in the preceding example. The point A at 22.75 psia represents the computed bubble-point pressure for a solution whose mole fraction of C_4H_{10} is 0.50. Point B represents the composition of the vapor at the bubble point. Similarly, the points C and D represent the bubble point and composition of the vapor at the bubble point for a solution whose mole fraction of C_4H_{10} is 0.75. The points E and F represent the vapor pressure of pure butane and pure propane, respectively, at 0° F. The line $FACE$ is the bubble-point line and the line $FBDE$ is the dew-point line. It is obvious that a pressure-composition diagram for any *ideal* binary system could be calculated in this manner and would serve to describe the phase behavior quantitatively.

It is of interest to note that the bubble-point line for an ideal binary solution is a linear function of composition. This follows from Raoult's Law since in this case the bubble-point pressure is given by

$$\text{BPP} = x_1P_1{}^0 + x_2P_2{}^0$$

Since $x_2 = 1 - x_1$ for a binary system, this equation becomes

$$\text{BPP} = x_1(P_1{}^0 - P_2{}^0) + P_2{}^0$$

which is linear in the composition x_1.

Calculation of the Liquid and Vapor Composition of a Two-Component System in the Two-Phase Region. As the pressure is reduced below the bubble point more and more vapor forms and the vapor becomes richer in the less volatile component. Consequently, for a system in the two-phase region the compositions of the liquid and the vapor are different and neither is equal to the overall composition of the system. A method for calculating the composition of the liquid and vapor in the two-phase region will now be presented. This method is applicable to binary systems only.

Consider a binary system in the two-phase region. If x_1 and x_2 represent the mole fractions of the two components in the liquid phase at pressure P_T, application of Raoult's Law leads to the equation

$$x_1P_1{}^0 + x_2P_2{}^0 = P_T \tag{4}$$

It should be emphasized that x_1 and x_2 in equation 4 represent the composition of the *liquid* and are in general not equal to the overall composition of the system. Since, for a two-component system

$$x_1 + x_2 = 1 \qquad \text{or} \qquad x_2 = 1 - x_1$$

x_2 in equation 4 may be eliminated and it is apparent that

$$x_1P_1{}^0 + (1 - x_1)P_2{}^0 = P_T \tag{5}$$

Solving for x_1 yields

$$x_1 = \frac{P_T - P_2^0}{P_1 - P_2^0}$$

If x_1 had been eliminated in equation 4 by substituting $x_1 = 1 - x_2$ the following expression for x_2 would have resulted

$$x_2 = \frac{P_T - P_1^0}{P_2^0 - P_1^0} = 1 - x_1 \qquad (6)$$

If Dalton's Law is applicable to the vapor

$$y_1 = \frac{P_1}{P_T} = \frac{x_1 P_1^0}{P_T} \qquad (7)$$

and

$$y_2 = \frac{P_2}{P_T} = \frac{x_2 P_2^0}{P_T} = 1 - y_1 \qquad (8)$$

Equations 5 to 8 may be used to calculate the composition of the liquid and the vapor in the two-phase region at pressure P_T, for a binary system. An example calculation illustrating the use of these equations is given below.

EXAMPLE. Assuming ideal solution behavior calculate the composition of the liquid and vapor at 180° F and 95 psia for a system containing one mole of *n*-butane and one mole of *n*-pentane.

The vapor pressures of the pure components at 180° F are $P_{C_4H_{10}}{}^0 = 160$ psia and $P_{C_5H_{12}}{}^0 = 54$ psia. If butane is designated as component 1 and pentane as component 2 solution of equation 5 for the mole fraction of C_4H_{10} in the liquid gives

$$x_{C_4H_{10}} = \frac{P_T - P_{C_5H_{12}}{}^0}{P_{C_4H_{10}}{}^0 - P_{C_5H_{12}}{}^0} = \frac{95 - 54}{160 - 54} = 0.394$$

To calculate the mole fraction of C_5H_{12} in the liquid equation 6 may be used. However, since $x_{C_4H_{10}} + x_{C_5H_{12}} = 1$ it is apparent that

$$x_{C_5H_{12}} = 1 - x_{C_4H_{10}} = 1 - 0.394 = 0.606$$

The mole fraction of C_4H_{10} in the vapor is computed by substitution in equation 7.

$$y_{C_4H_{10}} = \frac{x_{C_4H_{10}} P_{C_4H_{10}}{}^0}{P_T} = \frac{0.394 \times 160}{95} = 0.665$$

The mole fraction of C_5H_{12} in the vapor is

$$y_{C_5H_{12}} = 1 - y_{C_4H_{10}} = 1 - 0.665 = 0.335$$

The overall composition of the system does not appear in equations 5 to 8. The compositions of the liquid and vapor in the two-phase region are determined by the temperature and pressure *only*. This point has already been discussed in the preceding chapter. It will be recalled that the compositions of the liquid and vapor are represented by the extremities of a horizontal line through the two-phase region on a pressure-composition diagram. They are independent of the overall composition provided, of course, that the overall composition is such that the system exists as two phases at the temperature and pressure in question. However, the overall composition does determine the relative *amounts* of liquid and vapor in the system as is shown by equations 3 to 5 in Chapter 4.

Alternate Method for Calculating the Bubble-Point Pressure of an Ideal Two-Component System. Although Raoult's Law can be used directly to calculate the bubble-point pressure of an ideal solution, an alternate method which is applicable to two-component systems will now be presented. Since equations 5 to 8 are applicable anywhere in the two-phase region they apply at the bubble point and the dew point. At the bubble point the system is essentially all liquid except for an infinitesimal amount of vapor. Consequently, the composition of the liquid will be equal to the overall composition of the system. If the overall composition is substituted for x_1 and x_2 in equations 5 and 6 then *either* may be solved for P_T at a given temperature. The value of P_T calculated in this manner is equal to the bubble-point pressure. The *composition* of the infinitesimal amount of vapor at the bubble point may be computed by substitution in equations 7 and 8.

EXAMPLE. A system is composed of one mole of *n*-butane and one mole of *n*-pentane. Calculate the bubble-point pressure and the composition of the vapor at 180° F using the alternate method presented above.

Since the overall composition and the composition of the liquid are equal at the bubble point, the mole fraction of each component in the liquid is 0.50. At 180° F the vapor pressures of butane and pentane are 160 psia and 54 psia respectively. Substitution in equation 5 gives

$$x_{C_4H_{10}} = \frac{P_T - P_{C_5H_{12}}{}^0}{P_{C_4H_{10}}{}^0 - P_{C_5H_{12}}{}^0}$$

$$0.50 = \frac{P_T - 54}{160 - 54}$$

Solving for P_T gives

$$P_T = 107 \text{ psia} = \text{BPP}$$

The composition of the vapor at the bubble point is calculated using equation 7

$$y_{C_4H_{10}} = \frac{x_{C_4H_{10}} P_{C_4H_{10}}{}^0}{P_T} = \frac{0.5 \times 160}{107} = 0.747$$

and

$$y_{C_5H_{12}} = 1 - 0.747 = 0.253$$

Calculation of the Dew-Point Pressure for a Two-Component System. At the dew point the system is essentially all vapor except for an infinitesimal amount of liquid. Under these conditions the composition of the vapor is equal to the overall composition. According to equation 7

$$y_1 = \frac{x_1 P_1{}^0}{P_T}$$

Therefore if x_1, the mole fraction of component 1 in the liquid at the dew point, were known, its value could be substituted in equation 7 and the equation solved for the dew-point pressure P_T. However, x_1 is not known directly but according to equation 5

$$x_1 = \frac{P_T - P_2{}^0}{P_1{}^0 - P_2{}^0}$$

Substituting equation 5 in equation 7 it is apparent that

$$y_1 = \frac{(P_T - P_2{}^0/P_1{}^0 - P_2{}^0) \times P_1{}^0}{P_T} \tag{9}$$

This equation may be solved for the dew-point pressure P_T. Having calculated P_T, equations 5 and 6 may be used to compute the composition of the infinitesimal amount of liquid at the dew point.

EXAMPLE. Consider the same system as in the preceding examples and calculate the dew-point pressure and the composition of the liquid at the dew point at 180° F.

Since the overall composition and the composition of the vapor are equal at the dew point then the mole fraction of each component in the vapor is 0.50. Substitution in equation 9 gives

$$y_{C_4H_{10}} = \frac{(P_T - P_{C_5H_{12}}{}^0/P_{C_4H_{10}}{}^0 - P_{C_5H_{12}}{}^0) \times P_{C_4H_{10}}{}^0}{P_T}$$

$$0.50 = \frac{[(P_T - 54)/(160 - 54)]160}{P_T}$$

Solving for P_T, the dew-point pressure at 180° F, gives

$$P_T = 80.8 \text{ psia} = \text{DPP}$$

The composition of the liquid at the dew point is calculated using equation 3.

$$x_{C_4H_{10}} = \frac{P_T - P_{C_5H_{12}}{}^0}{P_{C_4H_{10}}{}^0 - P_{C_5H_{12}}{}^0} = \frac{80.8 - 54}{160 - 54} = 0.243$$

and

$$x_{C_5H_{12}} = 1 - 0.243 = 0.757$$

To summarize, the following quantities have been calculated in the *three preceding* examples for a system composed of one mole of *n*-butane and one mole of *n*-pentane at 180° F

1. Bubble-point pressure (107 psia)
2. Composition of vapor at the bubble point ($yC_4H_{10} = 0.747$, $yC_5H_{12} = 0.253$)
3. Composition of liquid and vapor at 95 psia ($xC_4H_{10} = 0.394$, $xC_5H_{12} = 0.606$, $yC_4H_{10} = 0.665$, $yC_5H_{12} = 0.335$)
4. Dew-point pressure (80.8 psia)
5. Composition of liquid at dew point ($xC_4H_{10} = 0.243$, $xC_5H_{12} = 0.757$)

These quantities can be represented on a pressure-composition diagram as shown in Figure 39. In this diagram the composition is expressed in terms of mole fraction of butane.

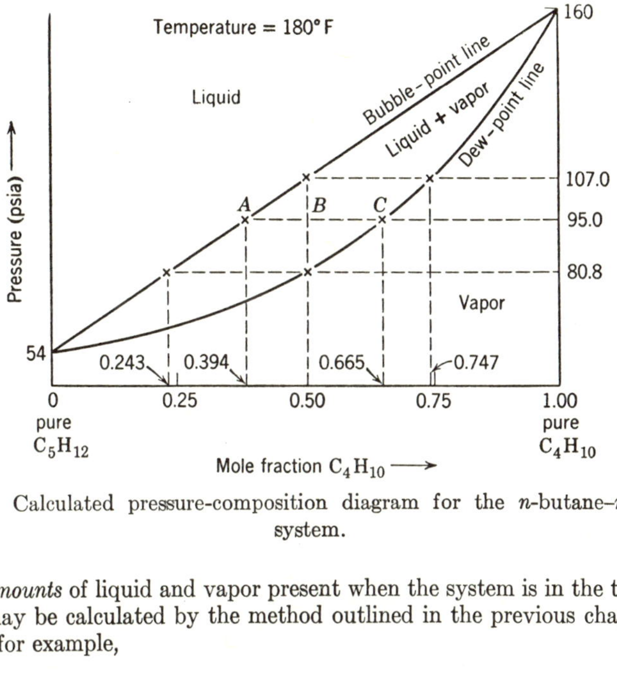

FIG. 39. Calculated pressure-composition diagram for the *n*-butane–*n*-pentane system.

The *amounts* of liquid and vapor present when the system is in the two-phase region may be calculated by the method outlined in the previous chapter. At 95 psia, for example,

$$\frac{\text{Moles of liquid}}{\text{Total moles}} = \frac{n_l}{n} = \frac{BC}{AC} = \frac{0.665 - 0.500}{0.665 - 0.394} = 0.609$$

Since $n = 2$, $n_l = 1.218$ moles and $n_v = 0.782$ mole. The apparent molecular weights of the liquid and vapor may be computed from their compositions using equation 11 in Chapter 2. The *weight* of liquid or vapor is equal to the product of its apparent molecular weight and the number of moles. That is,

$$\text{Weight of liquid} = \text{AMW}_l \times n_l$$

and

$$\text{Weight of vapor} = \text{AMW}_v \times n_v$$

Calculations Assuming Ideal Solution Behavior for Multicomponent Systems. The calculation of the bubble-point pressure and the composition of the vapor at the bubble point for an ideal solution consisting of more than two components involves no new principles or procedures. If Raoult's Law is applicable the partial pressure of each component in the vapor can be calculated and their sum is equal to the bubble-point pressure. Stated mathematically

$$\text{BPP} = \Sigma x_i P_i^0$$

where x_i is the mole fraction of the ith component in the liquid and P_i^0 is the vapor pressure of the pure ith component. Similarly, if Dalton's Law applies to the vapor phase the mole fraction of each component in the vapor is given by

$$y_i = \frac{P_i}{P_T} = \frac{x_i P_i^0}{\text{BPP}}$$

For a pressure at which partial evaporation occurs and both liquid and vapor are present in finite quantities the calculation of the composition of the liquid and the vapor in a multicomponent system is more complex but can be carried out in the following manner.

Let n = total number of moles in the system
n_l = total number of moles in the liquid
n_v = total number of moles in the vapor
z_i = mole fraction of ith component in the overall system
x_i = mole fraction of ith component in the liquid
y_i = mole fraction of ith component in the vapor

Then $z_i n$ = moles of ith component in the system
$x_i n_l$ = moles of ith component in the liquid
$y_i n_v$ = moles of ith component in the vapor

A material balance on the ith component leads to the following equation

$$z_i n = x_i n_l + y_i n_v \tag{10}$$

Application of Raoult's Law and Dalton's Law to the ith component gives

$$y_i = \frac{P_i}{P_T} = \frac{x_i P_i^0}{P_T} \tag{11}$$

Eliminating y_i with these two equations and solving for x_i yields

$$z_i n = x_i n_l + \frac{P_i^0}{P_T} x_i n_v$$

$$x_i = \frac{z_i n}{n_l + \dfrac{P_i^0}{P_T} n_v}$$

Since $\Sigma x_i = 1$ it follows that

$$\Sigma x_i = \Sigma \frac{z_i n}{n_l + \dfrac{P_i^0}{P_T} n_v} = 1 \qquad (12)$$

If x_i had been eliminated in equation 10 and the resulting equation solved for y_i the following equation would have been obtained.

$$\Sigma y_i = \Sigma \frac{z_i n}{n_v + n_l \dfrac{P_T}{P_i^0}} = 1 \qquad (13)$$

These two equations are equivalent and either one can be used to calculate the compositions of a two-phase multicomponent system. These equations are most readily solved by trial and error. To simplify the calculation, one mole of starting material is taken as a basis. In this case $n = 1$ and $n_l + n_v = 1$. A reasonable value of n_l or n_v is chosen and the required summation at the temperature and pressure in question is carried out using equation 12 or 13. If the sum is equal to one then each term in the sum is equal to x_i or y_i, depending on which equation was employed. If the summation does not equal one a second value of n_l or n_v must be chosen and the computation repeated. An example calculation illustrating this method is given below.

EXAMPLE. A system consists of 25 mole per cent propane, 30 mole per cent pentane and 45 mole per cent heptane at 150° F. Assuming ideal solution behavior calculate the composition of the liquid and the vapor at 20 psia.

Let $n = 1$ and assume $n_l = 0.45$. Then $n_v = 0.55$ and the following calculations are carried out according to equation 12.

(1)	(2)	(3)	(4)	(5)
Component	z_i	P_i^0	$\frac{P_i^0}{P_T} = \frac{P_i^0}{20}$	$x_i = \frac{z_i}{0.45 + (P_i^0/20)0.55}$
C_3H_8	0.25	345.0	17.25	0.025
C_5H_{12}	0.30	36.6	1.83	0.207
C_7H_{16}	0.45	5.0	0.25	0.766
				$\Sigma x_i = 0.998$

Since Σx_i is essentially equal to one the composition of the liquid is given in column 5. It is understood that if the sum had not been equal to one another value of n_l would be assumed. The composition of the vapor can be calculated using equation 13 and substituting $n = 1$, $n_l = 0.45$, and $n_v = 0.55$. However, it is simpler to apply equation 11 directly and, since x_i for each component is now known, the values of y_i may be computed. The composition of the vapor calculated in this manner is given below.

x_i	$y_i = \dfrac{x_i P_i^0}{P_T}$
0.025	$17.25 \times 0.025 = 0.431$
0.207	$1.83 \times 0.207 = 0.379$
0.766	$0.25 \times 0.766 = 0.191$

Calculations of the type just outlined are applicable in predicting the behavior of the fluids in an oil and gas separator, provided ideal solution behavior may be assumed. The fluid introduced into the separator is usually known as the feed stock. The vapor phase and the fluid phase are separated, the relative amounts and composition of each being determined by the composition of the feed stock and the operating temperature and pressure of the separator. If the solution is ideal it is necessary to know only the overall composition of the system and the vapor pressures of the pure components at the temperature in question in order to calculate the phase behavior. Indeed, for approximate calculations it is necessary to know only the composition of the feed stock and the boiling points of the pure components since the vapor pressure at any temperature may be estimated using Trouton's rule and the Clausius–Clapeyron equation.

NON-IDEAL SOLUTIONS

The Concept of Equilibrium Constants. If the solution is not ideal and Raoult's Law and Dalton's Law are not applicable it is necessary to make an empirical correction. For an ideal solution the relationship between the mole fractions of a given component in the liquid and vapor phases is given by

$$y_i = \frac{P_i^0}{P_T} x_i$$

For a non-ideal solution this relation becomes

$$y_i = K_i x_i$$

where K_i is an experimentally determined constant known as the equilibrium constant. In the case of an ideal solution the value of K_i is a function of temperature and pressure and is equal to the vapor pressure of the pure component divided by the total pressure. The value of K_i in a non-ideal solution is also a function of both temperature and pressure. Equilibrium constants increase with increasing temperature and decrease with increasing pressure. A typical plot of K_i as a function of temperature is shown in Figure 40. Curve 1 represents the value of K_i as a function of temperature at a pressure P_1 and curve 2 represents the value of K_i as a function of temperature at

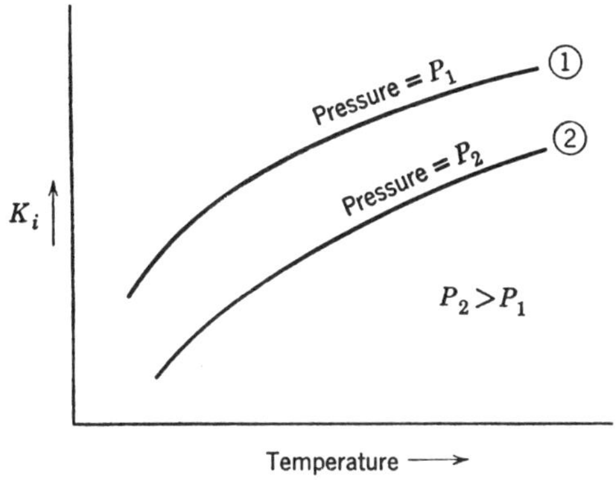

FIG. 40. Illustrative plot showing dependence of equilibrium constants on temperature and pressure.

pressure P_2 where P_2 is greater than P_1. Graphs of this type for a number of hydrocarbons are given in Figures 41 to 50. The use of these graphs is shown in the following example.

EXAMPLE. From the appropriate chart find the equilibrium constants for n-butane at 100° F and 25 psia and at 100° F and 400 psia. What would be the value of these constants if the solution containing the n-butane were ideal?

From Figure 45 the value of K_i is found by interpolation to be 2.10 at 100° F and 25 psia. At 100° F and 400 psia the value of K_i is 0.25.

Since the vapor pressure of butane is 52.2 psia at 100° F, the values of these constants for an ideal solution are

$$K_i \text{ (ideal, 25 psia)} = \frac{P_i^0}{P_T} = \frac{52.2}{25} = 2.09$$

and

$$K_i \text{ (ideal, 400 psia)} = \frac{52.2}{400} = 0.13$$

These results indicate that, for all practical purposes, at 100° F and 25 psia, the butane exhibits the behavior of an ideal solution component. However, at 400 psia the solution is not ideal insofar as the behavior of butane is concerned.

The values of these equilibrium constants have been determined experimentally. Since

$$K_i = \frac{y_i}{x_i}$$

the value of the equilibrium constant for the *i*th component is equal to the mole fraction of that component in the vapor divided by its mole fraction in the liquid. Consequently, to determine the value of the equilibrium constant at a given temperature and pressure the vapor phase and the liquid phase of a mixture of hydrocarbons are separated and carefully analyzed. These analyses are usually carried out by subjecting each phase to a careful fractional distillation. The mole fractions of each component in both the liquid and vapor are determined and the K values for each component are calculated at the temperature and pressure at which the separation was conducted. This process is repeated at a number of temperatures and pressures and graphs of K_i versus temperature at various pressures are constructed.

It is assumed that the K values of a given component are independent of the nature of the other components which constitute the solution. If the solution is ideal this assumption is exact. In the case of actual solutions it has been found that the K values become increasingly dependent on the overall composition as the pressure is increased. The upper limit of the pressure in Figures 41 to 50 is 800 psia but equilibrium constants have been determined at considerably higher pressures and may be found in the literature. However, to use these high-pressure equilibrium constants with confidence they should be employed in systems which are similar to the one used for the experimental determination of the constants. It has been found that the equilibrium constants given in Figures 41 to 50 are sufficiently accurate in the indicated pressure range for most engineering calculations, provided the system is not near the critical.

EXAMPLES OF THE USE OF EQUILIBRIUM CONSTANTS

Calculations of liquid and vapor compositions by means of equilibrium constants are, in general, quite similar to those previously outlined for ideal solutions. However, instead of using vapor pressures and the total pressure the K values are employed directly. It can readily be shown that equations 5 to 8 in terms of equilibrium constants become

$$x_1 = \frac{1 - K_2}{K_1 - K_2} \tag{14}$$

$$x_2 = 1 - x_1 = \frac{1 - K_1}{K_2 - K_1} \tag{15}$$

$$y_1 = K_1 x_1 \tag{16}$$

$$y_2 = 1 - y_1 = K_2 x_2 \tag{17}$$

These equations may be used to compute the liquid and vapor compositions of non-ideal *binary* solutions.

EXAMPLE. A two-component system contains one mole of *n*-butane and one mole of *n*-pentane. Calculate the composition of the liquid and the vapor at 180° F and 95 psia. Assume non-ideal solution behavior.

From the appropriate charts the K values are found to be

$$K_{C_4H_{10}} = 1.50, \qquad K_{C_5H_{12}} = 0.62$$

Substituting in equations 14 to 17 gives

$$x_{C_4H_{10}} = \frac{1 - K_{C_5H_{12}}}{K_{C_4H_{10}} - K_{C_5H_{12}}} = \frac{1 - 0.62}{1.50 - 0.62} = 0.431$$

$$x_{C_5H_{12}} = 1 - x_{C_4H_{10}} = 0.569$$

$$y_{C_4H_{10}} = K_{C_4H_{10}}x_{C_4H_{10}} = 1.50 \times 0.431 = 0.647$$

$$y_{C_5H_{12}} = 1 - y_{C_4H_{10}} = 0.353$$

In terms of equilibrium constants equations 12 and 13 for multicomponent systems become

$$\Sigma x_i = \Sigma \frac{z_i n}{n_l + K_i n_v} = 1 \tag{18}$$

and

$$\Sigma y_i = \Sigma \frac{z_i n}{n_v + n_l/K_i} = 1 \tag{19}$$

With these equations the composition of the liquid and the vapor in a non-ideal multicomponent system may be calculated by trial and error by essentially the same method that was employed to solve equations 12 and 13.

EXAMPLE. A hydrocarbon system has the following composition:

Component	Mole Fraction
CH_4	0.15
C_2H_6	0.05
C_3H_8	0.25
i-C_4H_{10}	0.05
n-C_4H_{10}	0.15
n-C_5H_{12}	0.25
n-C_6H_{14}	0.10

Calculate the composition of the liquid and the vapor if a separation is conducted at 200 psia and 100° F. Assume non-ideal solution behavior. For $n = 1$, assume

$n_l = 0.77$ and $n_v = 0.23$. Equation 18 is employed to make the following calculations.

(1) Component	(2) z_i	(3) K_i	(4) $x_i = \frac{z_i}{n_l + K_i n_v} = \frac{z_i}{0.77 + K_i \times 0.23}$	(5) $y_i = K_i x_i$
CH_4	0.15	14.1	0.037	0.522
C_2H_6	0.05	2.78	0.035	0.097
C_3H_8	0.25	0.97	0.252	0.245
$i\text{-}C_4H_{10}$	0.05	0.46	0.057	0.026
$n\text{-}C_4H_{10}$	0.15	0.35	0.177	0.062
$n\text{-}C_5H_{12}$	0.25	0.116	0.314	0.037
$n\text{-}C_6H_{14}$	0.10	0.041	0.128	0.005
			$\Sigma x_i = 1.000$	

The K values in column 3 are obtained from the appropriate chart at 200 psia and 100° F. Since $\Sigma x_i = 1.000$ the values in column 4 represent the composition of the liquid. It is understood that if $\Sigma x_i \neq 1$ another value of n_l must be chosen and the calculation repeated. The values in column 5 represent the composition of the vapor. To calculate the weight of liquid recovery the apparent molecular weight of the liquid is computed. The weight of liquid recovery per mole of starting material is given by the product of this apparent molecular weight and n_l. A similar calculation would yield the weight of vapor.

When $\Sigma x_i \neq 1$ the question arises whether to increase or decrease n_l for the next step in the trial-and-error solution. Since $n_v = 1 - n_l$ when $n = 1$ equation 18 becomes

$$\sum \frac{z_i}{n_l + K_i n_v} = \sum \frac{z_i}{n_l + K_i(1 - n_l)} = \sum \frac{z_i}{n_l(1 - K_i) + K_i} = 1$$

When the equation is written in this form it is apparent that if $\Sigma x_i > 1$ the assumed value of n_l should be increased and if $\Sigma x_i < 1$ n_l should be decreased.

Calculation of Bubble-Point Pressure and Dew-Point Pressure Using Equilibrium Constants. Since the total pressure P_T no longer appears implicitly in equations 14 to 19 but is contained in the K values, it is no longer possible to solve directly for the bubble-point and dew-point pressure as was done in the case of ideal solutions. A method will now be presented for calculating the bubble-point pressure and the dew-point pressure, which is applicable to *both* binary and multicomponent systems which are non-ideal. At the bubble point the system is entirely in the liquid state except for an infinitesimal amount of vapor. Consequently, since $n_v = 0$ and $n = n_l$ equation 19 becomes

$$1 = \Sigma \frac{z_i n}{n_v + n_l/K_i} = \Sigma \frac{z_i n}{0 + n/K_i} = \Sigma K_i z_i \tag{20}$$

This equation states that at the bubble point the sum of the products of the equilibrium constant and the mole fraction of the component in

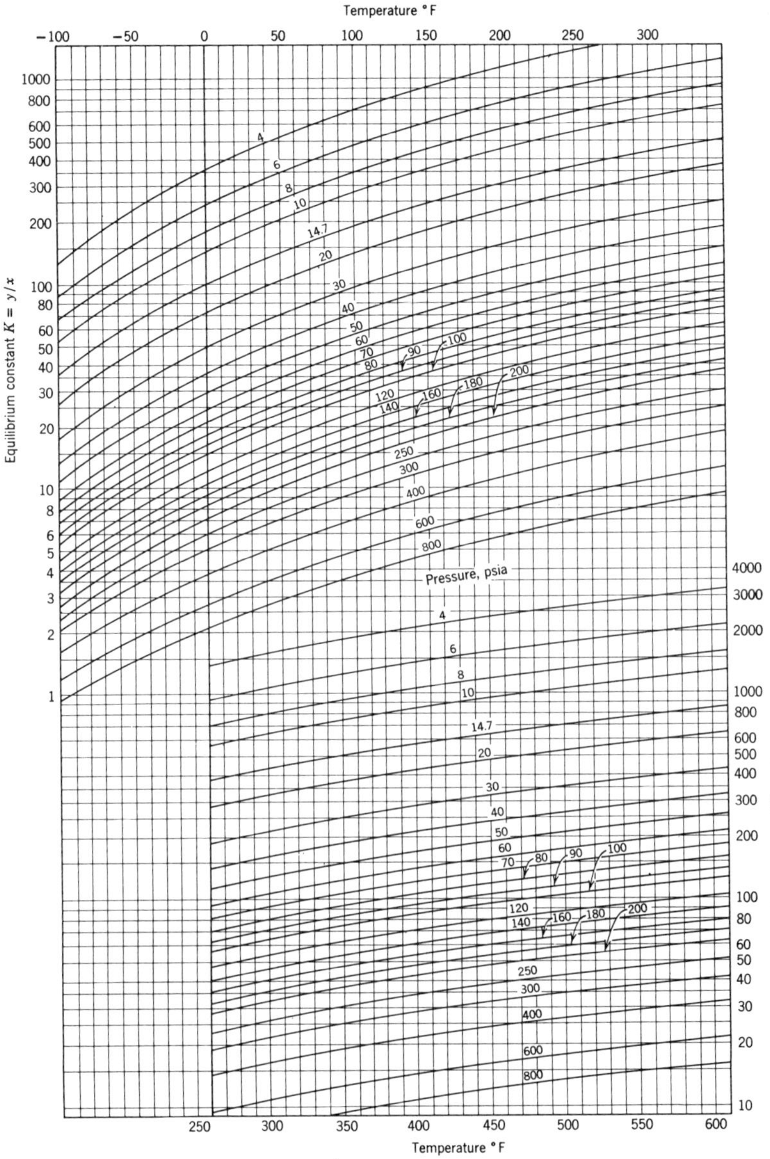

Fig. 41. Equilibrium constants for methane. (Natural Gasoline Supply Men's Association, *Engineering Data Book,* 1951, p. 106.)

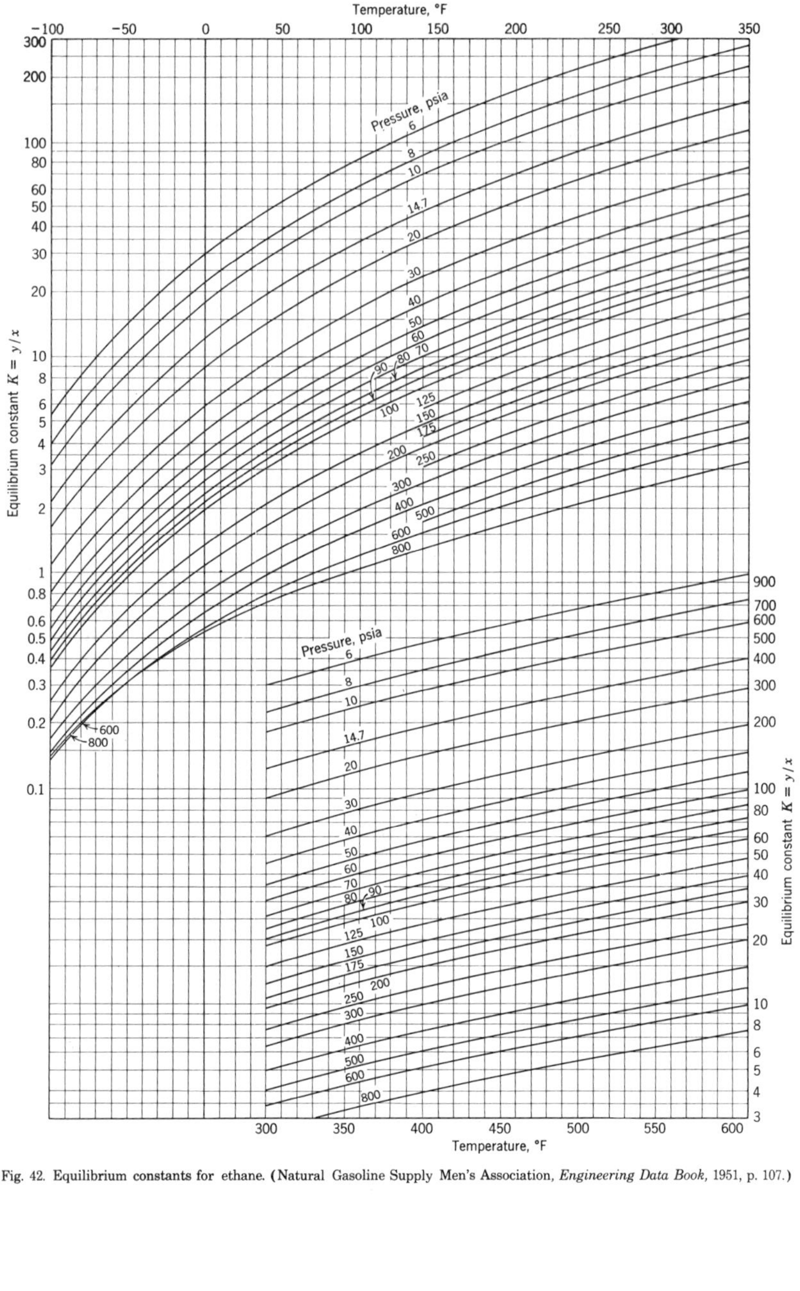

Fig. 42. Equilibrium constants for ethane. (Natural Gasoline Supply Men's Association, *Engineering Data Book*, 1951, p. 107.)

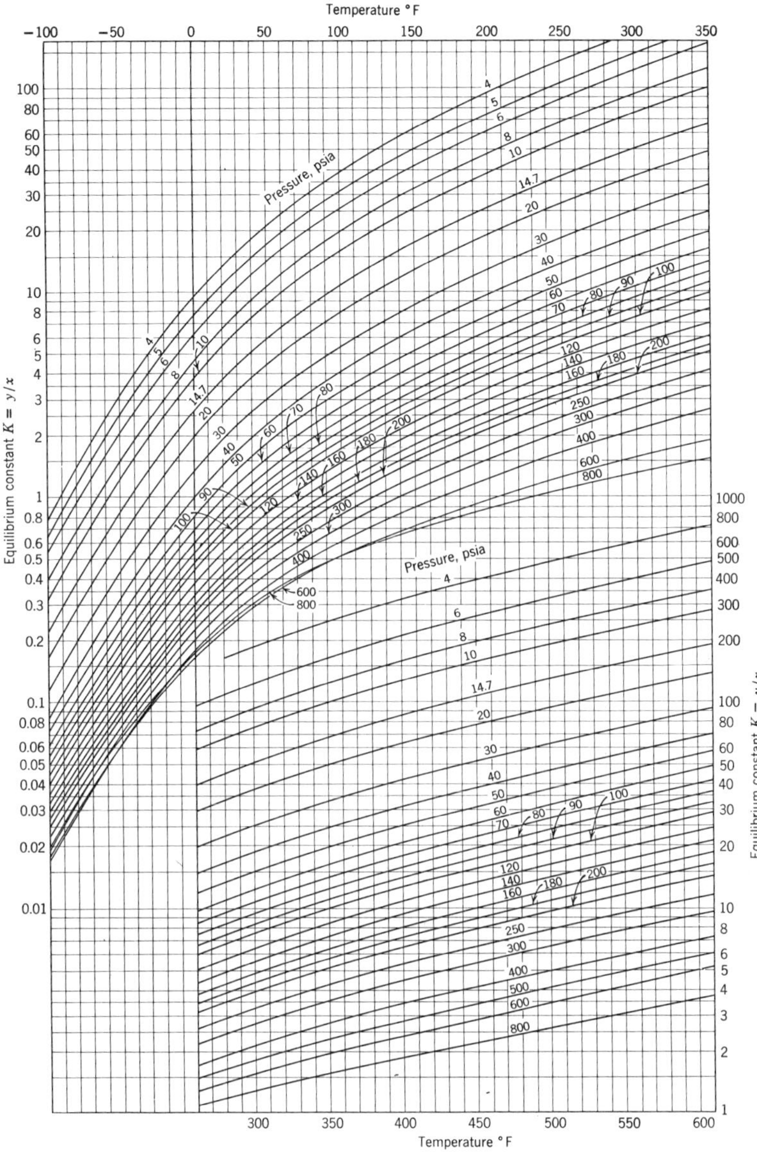

Fig. 43. Equilibrium constants for propane. (Natural Gasoline Supply Men's Association, *Engineering Data Book,* 1951, p. 108.)

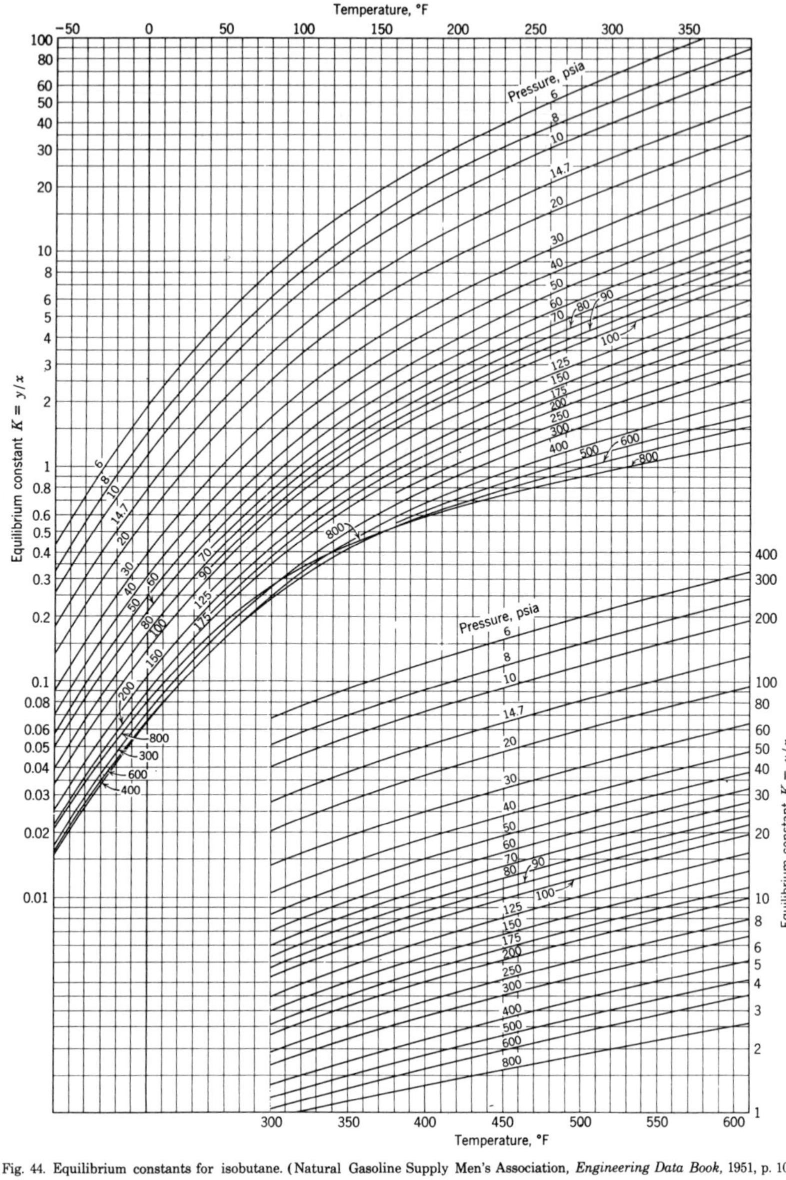

Fig. 44. Equilibrium constants for isobutane. (Natural Gasoline Supply Men's Association, *Engineering Data Book*, 1951, p. 109.)

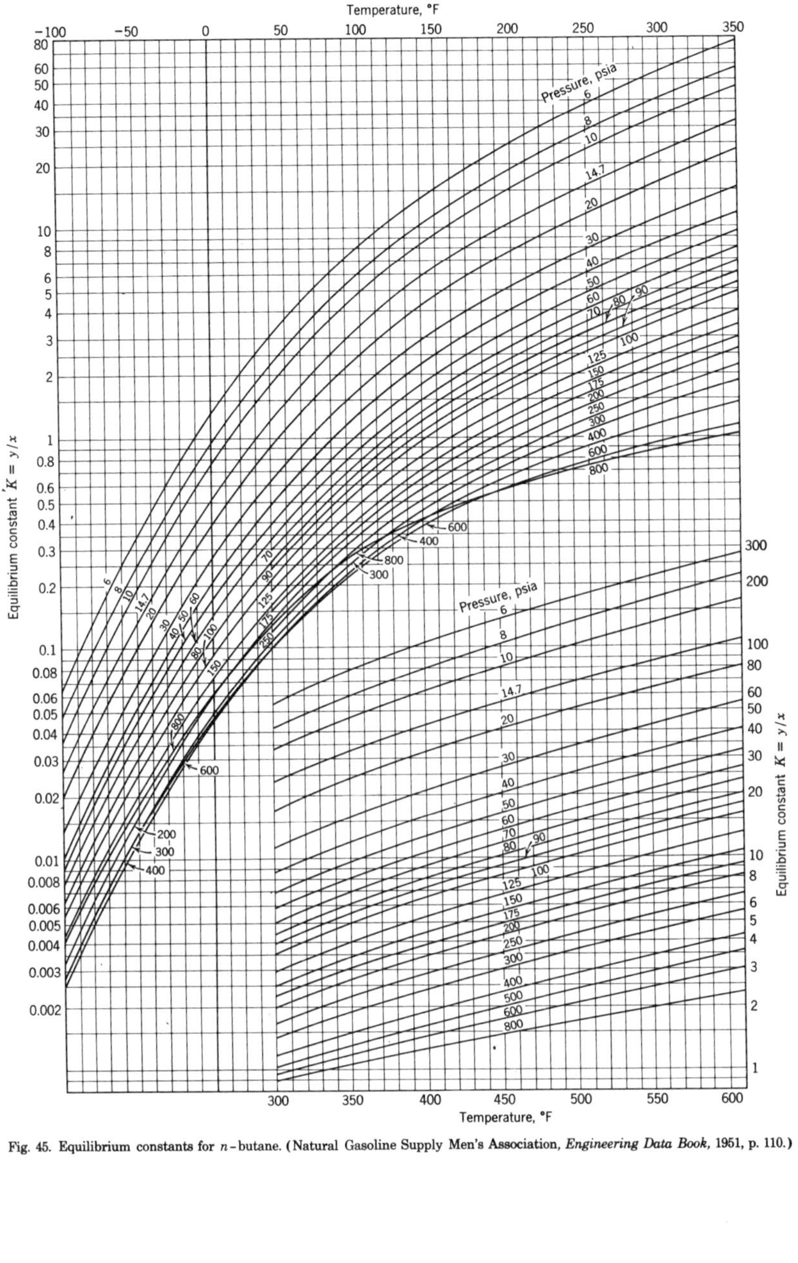

Fig. 45. Equilibrium constants for n-butane. (Natural Gasoline Supply Men's Association, *Engineering Data Book*, 1951, p. 110.)

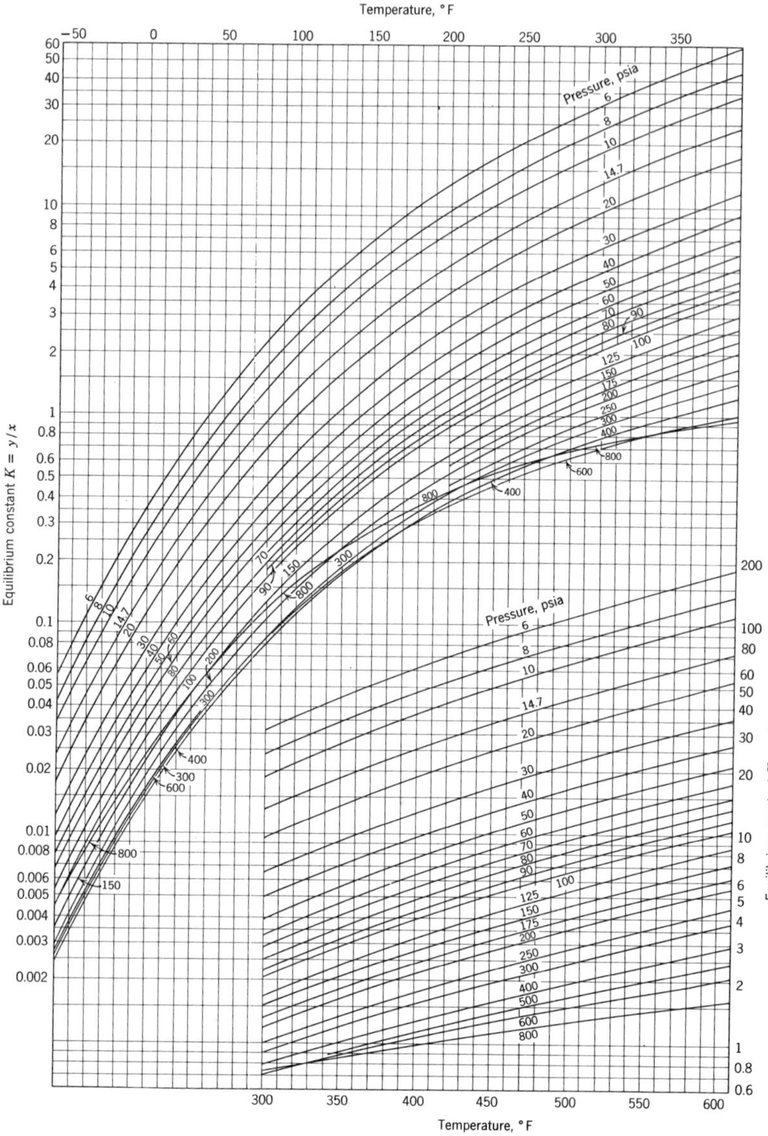

Fig. 46. Equilibrium constants for isopentane. (Natural Gasoline Supply Men's Association, *Engineering Data Book,* 1951, p. 111.)

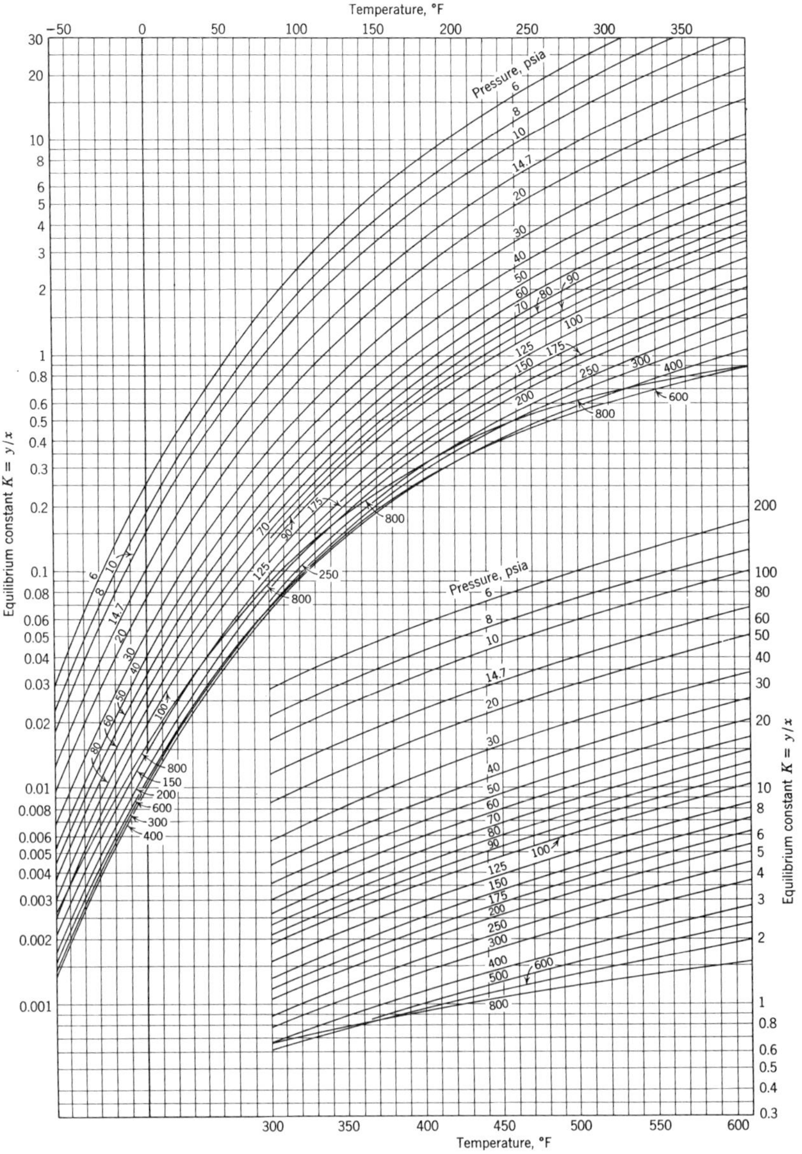

Fig. 47. Equilibrium constants for *n*-pentane. (Natural Gasoline Supply Men's Association, *Engineering Data Book*, 1951, p. 112.)

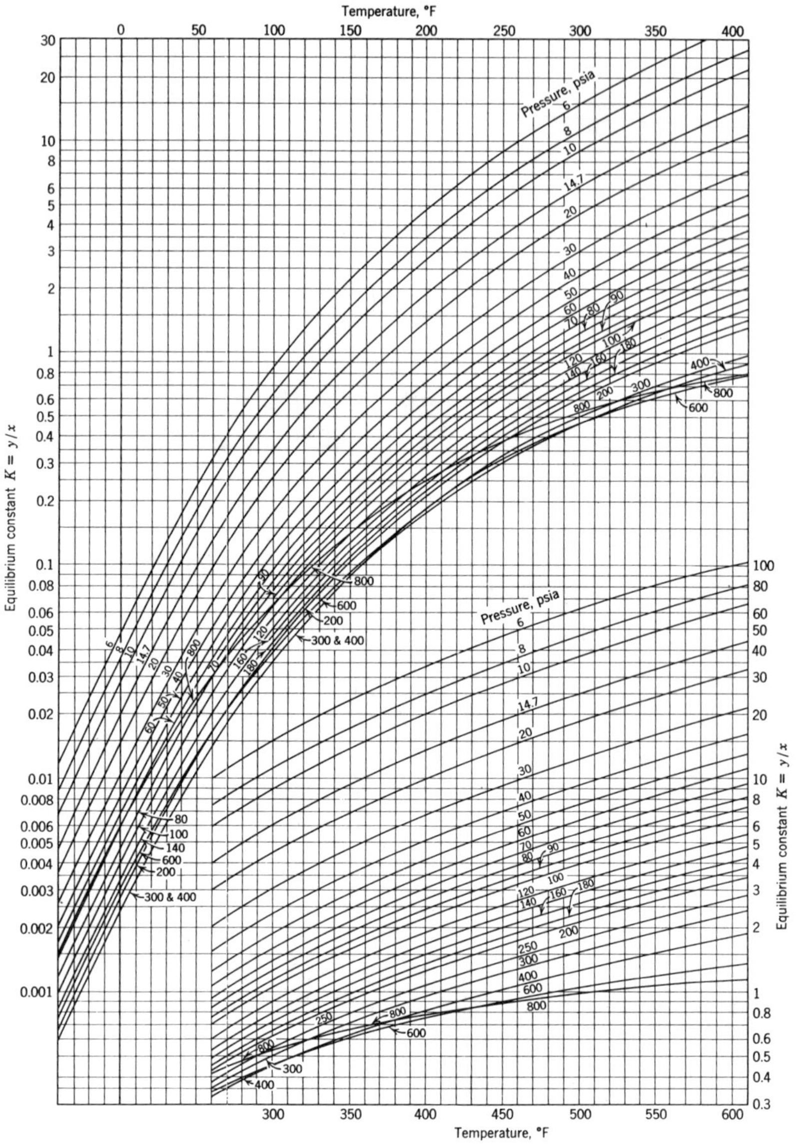

Fig. 48. Equilibrium constants for *n*-hexane. (Natural Gasoline Supply Men's Association, *Engineering Data Book*, 1951, p. 113.)

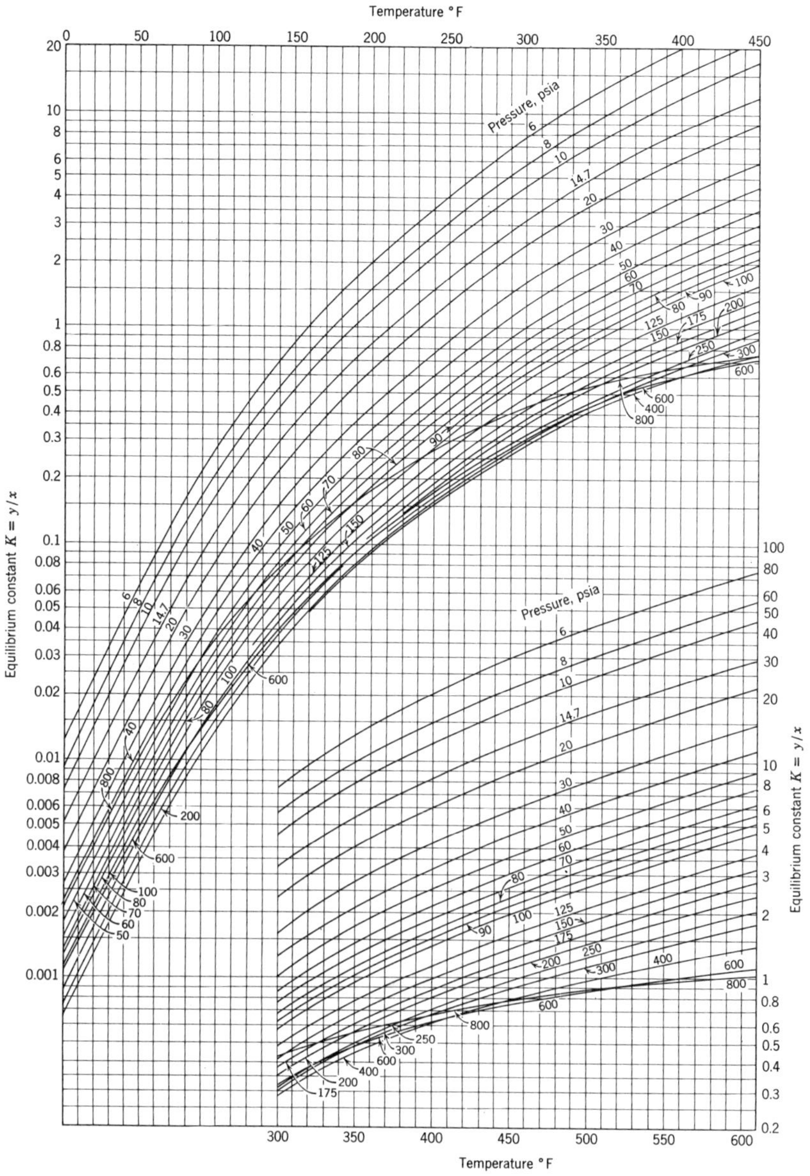

Fig. 49. Equilibrium constants for n-heptane. (Natural Gasoline Supply Men's Association, *Engineering Data Book*, 1951, p. 114.)

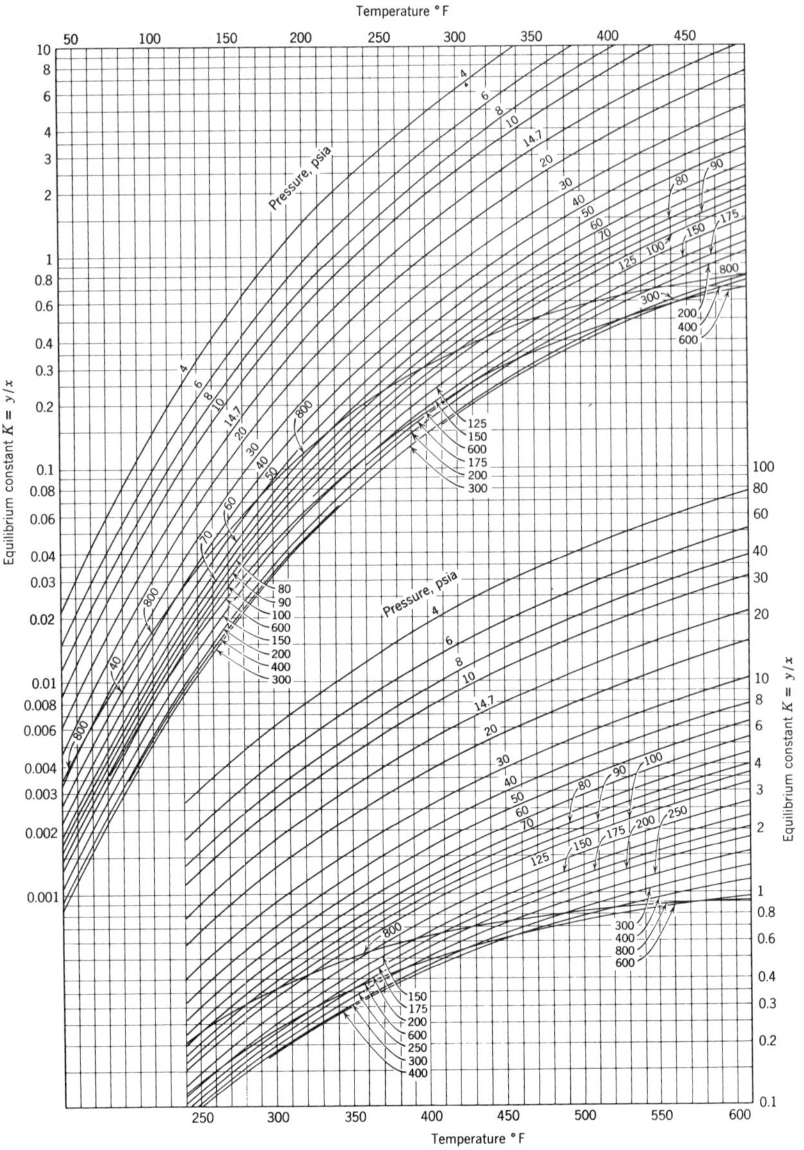

Fig. 50. Equilibrium constants for n-octane. (Natural Gasoline Supply Men's Association, *Engineering Data Book*, 1951, p. 115.)

the entire system must equal one. The values of z_i represent the mole fractions in the entire system but since the system is essentially all liquid at the bubble point they also represent the mole fractions in the liquid phase. For this reason equation 20 is often written in the form

$$\Sigma K_i x_i = 1 \tag{21}$$

To calculate the bubble-point pressure at a given temperature using this equation it is necessary to choose a pressure and carry out the summation as indicated. An approximate bubble-point pressure calculated using Raoult's Law in the manner previously outlined for ideal solutions may be used as a point of departure. If the sum is less than one the process is repeated at a lower pressure. Conversely, if the sum is greater than one a higher pressure is chosen. Interpolation may be employed to determine the bubble-point pressure when it has been bracketed between two values of pressure, one of which gives a value of $\Sigma K_i x_i$ slightly greater than one and the other gives a value of $\Sigma K_i x_i$ slightly less than one. When the bubble-point pressure has been established by this trial-and-error process the composition of the infinitesimal amount of vapor at the bubble point is given by the individual products in $\Sigma K_i x_i$ computed at the bubble-point pressure.

To calculate the dew-point pressure using equilibrium constants a similar procedure is carried out. At the dew point the system is entirely in the vapor state except for an infinitesimal amount of liquid. Consequently, since $n_l = 0$ and $n = n_v$ equation 18 becomes

$$1 = \sum \frac{z_i n}{n_l + K_i n_v} = \sum \frac{z_i}{K_i} = \sum \frac{y_i}{K_i} \tag{22}$$

The dew-point pressure is the pressure at which the above summation is equal to one. This pressure is again found by a trial-and-error process. Here again the composition of the infinitesimal amount of liquid at the dew point may be computed by carrying out the summation at the dew-point pressure. The individual quotients obtained in this manner represent the composition of the liquid at the dew point.

Example. A system has the following overall composition

Component	Mole Fraction
n-C_4H_{10}	0.403
n-C_5H_{12}	0.325
n-C_6H_{14}	0.272

At 160° F calculate the bubble-point pressure, the composition of the vapor at the bubble point, the dew-point pressure, and the composition of the liquid at the dew point.

By Raoult's Law the approximate bubble-point pressure is $\Sigma x_i P_i^0$. At 160° F this becomes

$$\text{BPP} = 0.403 \times 123 + 0.325 \times 43.0 + 0.272 \times 15.8 = 67.85 \text{ psia}$$

Consequently, the first trial-and-error solution using equation 21 is made at 70 psia

Component	x_i	K_i (70 psia and 160° F)	$K_i x_i$
$n\text{-}C_4H_{10}$	0.403	1.63	0.657
$n\text{-}C_5H_{12}$	0.325	0.61	0.198
$n\text{-}C_6H_{14}$	0.272	0.25	0.068
			0.923

Since the summation is less than one a lower pressure of 60 psia is chosen

Component	x_i	K_i (60 psia and 160° F)	$K_i x_i$
$n\text{-}C_4H_{10}$	0.403	1.86	0.750
$n\text{-}C_5H_{12}$	0.325	0.70	0.228
$n\text{-}C_6H_{14}$	0.272	0.285	0.077
			1.055

The summation is greater than one so evidently the bubble-point pressure is between 60 and 70 psia. By interpolation the bubble point is found to be

$$\frac{70 - 60}{70 - \text{BPP}} = \frac{0.923 - 1.055}{0.923 - 1.000} \qquad \text{BPP} = 64.2 \text{ psia}$$

This interpolation can also be carried out graphically. A plot of $\Sigma K_i x_i$ versus pressure is constructed as shown in Figure 51. A linear relation is assumed, provided the pressure range is not too great and the pressure at which $\Sigma K_i x_i = 1$ can be read directly from the graph.

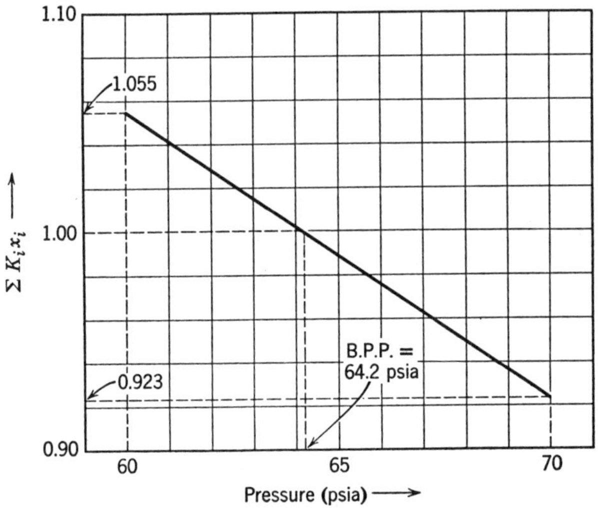

Fig. 51. Graphical interpolation of bubble-point pressure.

To calculate the composition of the vapor at the bubble point the K values at 160° F and 64.2 psia are obtained by interpolation from the appropriate charts and $K_i x_i$ computed.

Component	x_i	K_i (64.2 psia and 160° F)	$y_i = K_i x_i$
n-C_4H_{10}	0.403	1.76	0.709
n-C_5H_{12}	0.325	0.66	0.214
n-C_6H_{14}	0.272	0.27	0.073
			0.996

The values in the last column represent the vapor composition.

To calculate the dew-point pressure choose a pressure of 35 psia as a point of departure and calculate $\Sigma \frac{y_i}{K_i}$ as required by equation 22.

Component	y_i	K_i (35 psia and 160° F)	y_i/K_i
n-C_4H_{10}	0.403	3.17	0.127
n-C_5H_{12}	0.325	1.17	0.278
n-C_6H_{14}	0.272	0.47	0.579
			0.984

Since the required summation is too low the calculation is repeated at 40 psia

Component	y_i	K_i (40 psia and 160° F)	y_i/K_i
n-C_4H_{10}	0.403	2.75	0.147
n-C_5H_{12}	0.325	1.01	0.322
n-C_6H_{14}	0.272	0.41	0.663
			1.132

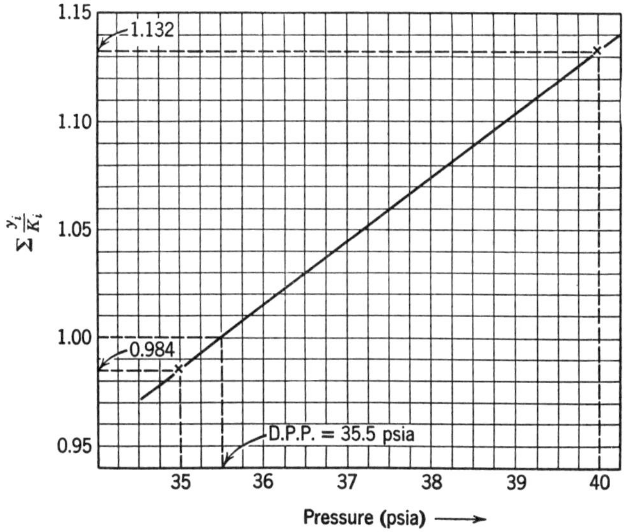

FIG. 52. Graphical interpolation of the dew-point pressure.

Since this summation is greater than one the dew-point pressure is between 35 psia and 40 psia. Interpolation leads to a value of 35.5 psia, as is shown in Figure 52. The composition of the dew-point liquid is computed as shown below

Component	y_i	K_i (35.5 psia and 160° F)	$x_i = \frac{y_i}{K_i}$
n-C_4H_{10}	0.403	3.13	0.129
n-C_5H_{12}	0.325	1.15	0.283
n-C_6H_{14}	0.272	0.464	0.586
			0.998

Henry's Law. Henry's Law expresses the effect of pressure on the solubility of a gas in a liquid. It may be stated: The weight of gas dissolved in a given quantity of liquid and at a given temperature is directly proportional to the pressure of the gas.

Henry's Law and Raoult's Law are related in the following manner. Equation 1 is applicable to any component in an ideal solution. However, if the component in question is a gas dissolved in a liquid it is unlikely that the solution will exhibit ideal behavior. Consequently, under these conditions, equation 1 becomes

$$P_i = Cx_i \tag{23}$$

where the proportionality constant C is an experimentally determined constant and not the vapor pressure of the pure component. If the liquid is relatively non-volatile P_i is essentially equal to P_T and, if the gas is not too soluble, x_i is proportional to W, the weight of gas dissolved. Under these conditions equation 23 becomes

$$P_T = C'W$$

where C' is the new proportionality constant and W is the weight of gas dissolved. This equation will be recognized as a mathematical statement of Henry's Law as given above.

Henry's Law is not an exact law. At high pressures and in cases where the gas is very soluble in the liquid marked deviations occur. As will be shown in the next chapter, solution of natural gases in crude oil do not follow Henry's Law except at very low pressures.

REFERENCES

Calhoun, J., *Fundamentals of Reservoir Engineering,* University of Oklahoma Press, Norman, Okla. (1953).

Daniels, F., *Outlines of Physical Chemistry,* John Wiley & Sons, New York (1948).

Huntington, R. L., *Natural Gas and Natural Gasoline,* McGraw-Hill Book Co., New York (1950).

Muskat, M., *Physical Principles of Oil Production,* McGraw-Hill Book Co., New York (1949).

Nelson, W. L., *Petroleum Refining Engineering,* McGraw-Hill Book Co., New York (1936).

Pirson, S., *Elements of Oil Reservoir Engineering,* McGraw-Hill Book Co., New York (1950).

PROBLEMS

1. What is the mole fraction of H_2O in a solution of 46 grams of ethyl alcohol in 100 grams of H_2O? What are the percentages by weight?

2. One pound-mole of *n*-butane and one pound-mole of *n*-pentane are mixed at 0° F. Assuming ideal-solution behavior, what is the volume of the resulting solution? The specific volumes of butane and pentane are 0.0259 cu ft/lb and 0.0243 cu ft/lb, respectively. What is the bubble-point pressure?

3. Calculate the bubble-point pressure and the composition of the vapor for a solution containing 100 lb of propane and 90 lb of *n*-pentane at 100° F. Assume ideal-solution behavior.

4. Construct a pressure versus composition diagram for *n*-hexane and *n*-heptane at 150° F.

5. An ideal solution containing one pound-mole propane and one pound-mole butane is heated to 100° F.

(*a*) What is the bubble-point pressure? *Answer:* 120 psia.

(*b*) What is the composition of the vapor at the bubble point? *Answer:* $y_{C_3H_8} = 0.783$.

(*c*) What is the dew-point pressure? *Answer:* 81.6 psia.

(*d*) What is the composition of the liquid at dew point? *Answer:* $x_{C_3H_8} = 0.218$.

(*e*) At a pressure midway between the dew-point pressure and the bubble-point pressure calculate the composition of the liquid and vapor.

(*f*) At a pressure midway between the dew-point pressure and the bubble-point pressure calculate the weights of the liquid phase and the vapor phase in pounds.

6. An ideal solution contains 7 pound-moles of *n*-butane and 3 pound-moles of pentane. Calculate the dew-point pressure at 180° F and the composition of the liquid at the dew point.

7. What is the composition of liquid and vapor of a solution of one mole of *n*-pentane and 2 moles of *n*-hexane at 140° F and one standard atmosphere pressure?

8. An ideal solution of *n*-butane and pentane at 180° F has a mole fraction of *n*-butane equal to 0.15 in the overall system. Calculate the composition of the liquid and the vapor at 110 psia and 180° F. What do these results indicate about the phase state of this system?

9. Calculate the bubble-point pressure at 120° F for the following mixture. Assume ideal-solution behavior.

Component	*Wt*
C_3H_8	5 lb
C_4H_{10}	30 lb
C_5H_{12}	10 lb
C_6H_{14}	30 lb
C_7H_{16}	25 lb

Answer: BPP = 51.3 psia.

10. Two hydrocarbons, A and B, form an ideal solution. The heats of vaporization are 9630 and 11,000 Btu per pound-mole, respectively. The normal boiling points are 31° F and 97° F, respectively. For a system containing 2 pound-moles of A and 3 pound-moles of B, calculate the bubble-point pressure at 97° F *Answer:* 27.7 psia.

11. Show that

$$\sum \frac{z_i n}{n_v + n_l \dfrac{P_T}{P_i^0}} = 1$$

12. From tables what are the values of following equilibrium constants: (*a*) Propane at 200° F and 100 psia; (*b*) butane at 100° F and 250 psia; (*c*) pentane at 150° F and 22 psia.

13. Using Figure 48 make a cross plot of the equilibrium constant data for n-hexane. Plot log K versus log P at 50° F, 150° F, and 250° F. Plot log K versus log P at the same temperature assuming ideal-solution behavior.

14. A solution of one mole of butane and one mole of propane is heated to 100° F and 103.4 psia. Using equilibrium constants calculate the composition of the vapor and the liquid. *Answer:* $x_{C_3H_8} = 0.408$ $y_{C_3H_8} = 0.675$.

15. A hydrocarbon mixture is made up of 5 pound-moles of n-hexane and 5 pound-moles of *iso*-butane. It exists at 200 psia and 200° F. Calculate the amount and composition of liquid recovery per mole of starting material if it is separated at 60 psia and 200° F. Calculate the same quantities if it is separated at 90 psia and 200° F and then if the liquid is again separated at 60 psia and 200° F. Use equilibrium constants.

16. A system contains 25 mole per cent propane, 30 mole per cent pentane and 45 mole per cent heptane at 150° F. Using equilibrium constants calculate the composition of the liquid and vapor at 20 psia. (*Hint:* Assume $n_l = 0.46$ p[illegible] mole of starting material.) What is the weight of liquid obtained per mole [illegible] starting material?

17. A solution composed of one mole of ethane and four moles of butane [illegible] non-ideal at 100° F. Calculate the bubble-point pressure.

18. Using equilibrium constants calculate the bubble-point and dew-point pr[illegible]sures at 120° F for the hydrocarbon system described in Problem 9. *Answer*[illegible] BPP = 48 psia.

19. The solubility of ethane in water is 0.0725 gram per liter at 60° F and one atm pressure. Calculate the solubility in standard cu ft per barrel at 400 psia if Henry's Law applies.

CHAPTER 6

RESERVOIR FLUID CHARACTERISTICS

For reservoir engineering calculations various properties of the crude oil and its associated gas and water must be known. It will be shown that theoretically many of these properties could be calculated by the methods presented in previous chapters, provided the composition of the system is known and complete equilibrium constant data for all of the components are available. However, since this information is seldom at hand, values of the reservoir fluid characteristics are usually experimentally determined or approximated by methods that experience has shown to be sufficiently accurate for most engineering computations.

This chapter will deal with both the hydrocarbon fluid characteristics and the "connate" or "interstitial" water characteristics. These characteristics will be defined, and methods for their computation or estimation presented. Also the laboratory procedures commonly employed in their experimental determination will be outlined.

THE HYDROCARBON FLUID CHARACTERISTICS

The hydrocarbon fluid characteristics that will be described in this chapter are listed below.

1. Gas-Formation Volume Factor (v). The gas-formation volume factor is defined as the volume in barrels occupied by one standard cubic foot of gas when subjected to reservoir temperature and pressure.

2. Gas Solubility (r). This represents the solubility of gas in crude oil at a given reservoir temperature and pressure measured in standard cubic feet per stock tank barrel of oil, that is, per barrel of oil measured at 60° F and 14.7 psia.

3. Oil-Formation Volume Factor (β). This denotes the volume at reservoir conditions occupied by one stock tank barrel of oil plus the gas in solution.

4. Two-Phase Formation Volume Factor (u). This is defined as the volume occupied in the reservoir at a given pressure by one stock tank barrel of oil plus the free gas which was initially dissolved in it.

5. Reservoir Fluid Viscosities (μ_g *and* μ_o). These viscosities represent the viscosity of the gas and of the oil under reservoir conditions.

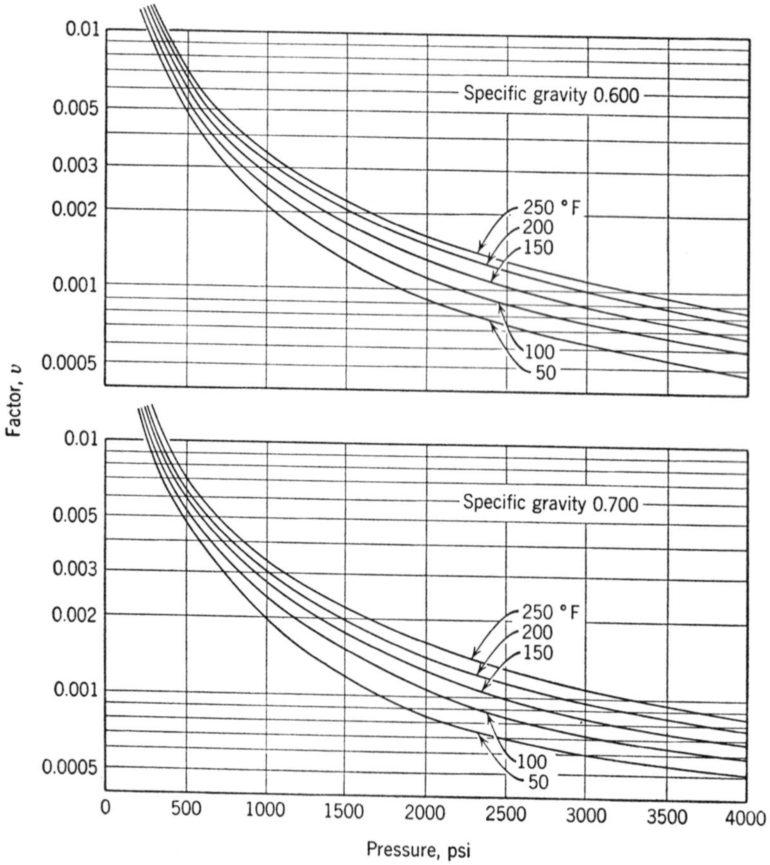

FIG. 53. The gas-formation volume factor as a function of temperature and pressure for various gravity gases. (From Calhoun, John C., Jr., *Fundamentals of Reservoir Engineering,* Copyright 1953, University of Oklahoma Press, p. 19.)

The Gas-Formation Volume Factor. The gas laws discussed in Chapter 2 may be employed to calculate the number of barrels of reservoir space occupied by one standard cubic foot of gas. If one standard cubic foot of gas is placed in a reservoir at reservoir pressure P_0 and temperature T_0 the following equation is valid provided the gas remains in the gaseous state.

$$\frac{P_1 V_1}{Z_1 T_1} = \frac{P_0 V_0}{Z_0 T_0}$$

The subscript 1 denotes standard conditions (i.e., 14.7 psia and 60° F) and the subscript 0 refers to reservoir conditions. Solving for V_0,

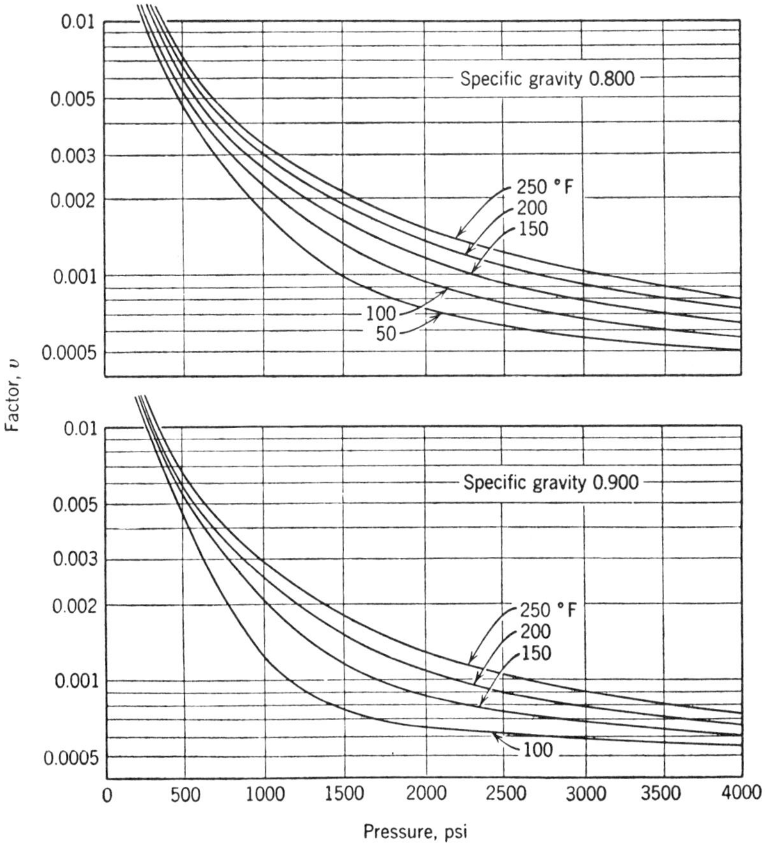

FIG. 53 (*Continued*).

keeping in mind that $Z_1 = 1$ for all practical purposes, it is apparent that

$$V_0 = \frac{Z_0 P_1 V_1 T_0}{Z_1 P_0 T_1} = \frac{Z_0 \times 14.7 \times 1 \times T_0}{1 \times P_0 \times 520} = 0.0283 \frac{Z_0 T_0}{P_0}$$

V_0 calculated in this manner will be in cubic feet. To express this volume in barrels it is necessary to divide by the factor 5.62 which represents the number of cubic feet in a barrel. Therefore, the gas-formation volume factor is given by

$$v = \frac{V_0}{5.62} = \frac{0.0283 Z_0 T_0}{5.62 P_0} = 0.00504 \frac{Z_0 T_0}{P_0} \text{ barrels/S.C.F.} \tag{1}$$

To compute v at a given reservoir temperature and pressure the value of the compressibility factor under these conditions must be known. If an experimental value of Z is not available and it is necessary to estimate a value for a reservoir gas one has recourse to the methods described in Chapter 2. If the composition of the gas is known a pseudo-reduced temperature and pressure may be calculated and the compressibility factor obtained from Figure 10. If, on the other hand, the composition is not known but a value of the gas gravity is available, it is still possible to evaluate the pseudo-critical temperature and pressure from Figures 11 and 12. With these pseudo-criticals and the values of the reservoir temperature and pressure, the pseudo-reduced temperature and pressure can be computed and the compressibility factor obtained from Figure 10 as before.

Example. A gas has a specific gravity of 0.740. Calculate v at 210° F and 2300 psia. From Figures 11 and 12 a gas whose specific gravity is 0.740 has the following pseudo-critical temperature and pressure

$$\underline{T_c} = 396 \qquad \underline{P_c} = 665$$

At 210° F and 2300 psia the pseudo-reduced temperature and pressure are

$$T_R = \frac{460 + 210}{396} = 1.69 \qquad P_R = \frac{2300}{665} = 3.46$$

At this reduced temperature and pressure the value of Z is 0.86 from Figure 10. Therefore, according to equation 1

$$v = 0.00504 \frac{Z_0 T_0}{P_0} = \frac{0.00504 \times 0.86 \times 670}{2300} = 0.00126 \text{ barrels/S.C.F.}$$

Values of v calculated in this manner for gases of various specific gravity are shown in Figure 53. In these charts the value of v is plotted as a function of reservoir pressure and temperature for gases with gravities of 0.60, 0.70, 0.80, and 0.90. For other gas gravities values of v may be obtained by interpolation with sufficient accuracy for most engineering computations.

Gas Solubility. The solubility of natural gas in oil must be referred to some basis and for this purpose it is customary to use one barrel of stock tank oil. The gas solubility (r) is defined as the number of cubic feet of gas measured at standard conditions which are in solution in one barrel of stock tank oil at reservoir temperature and pressure. A typical gas solubility curve as a function of pressure is shown in Figure 54 for a saturated crude oil at reservoir temperature.

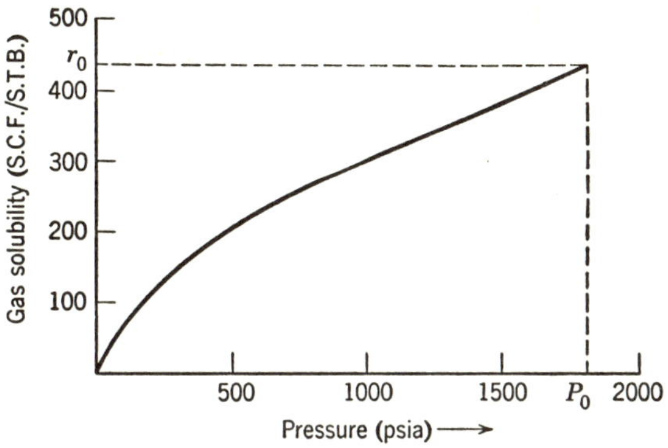

FIG. 54. Typical gas-solubility curve as a function of pressure for a saturated crude oil. Gas solubility is expressed in standard cubic feet per stock tank barrel of oil.

P_0 denotes the original reservoir pressure and r_0 is the original value of the gas solubility. As the pressure is reduced solution gas is liberated and the value of r decreases as shown. If the original crude is undersaturated at the initial reservoir temperature and pressure a reduction in pressure to the bubble point or saturation pressure is necessary before gas is evolved from the solution. A typical gas solubility curve for an undersaturated crude is shown in Figure 55. P_0 and P_s represent the original reservoir pressure and the saturation pressure, respectively. Between P_0 and P_s the gas solubility remains constant at r_0 but at pressures below P_s gas is evolved and r decreases as shown.

If the pressure is released from a sample of reservoir crude oil the quantity of gas evolved depends upon the conditions of liberation. There are two basic types of gas liberation: flash and differential. In a flash liberation the pressure is reduced by a finite amount and after equilibrium is established the gas is bled off, keeping the pressure constant. In a differential liberation the gas evolved is removed continuously from contact with the oil. The liquid is in equilibrium only with the gas being evolved at a given pressure and not with the gas

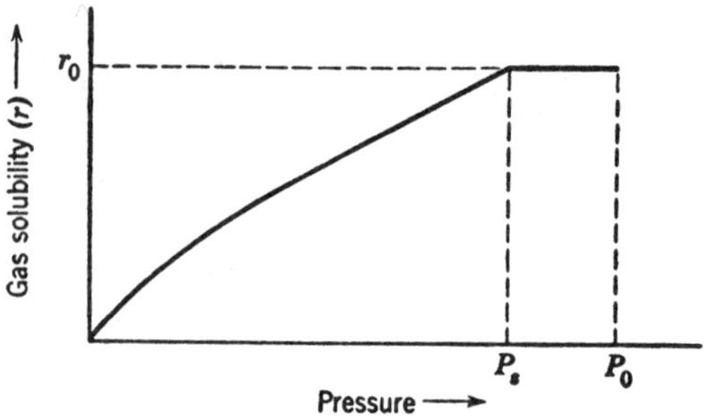

FIG. 55. Illustrative plot of r as a function of pressure for an undersaturated crude oil.

evolved over a finite pressure range. It is apparent that a series of flash liberations with infinitely small pressure reductions approaches a differential liberation. The two methods of liberation give different results for r, as is shown in Figure 56, the values of r for flash liberation

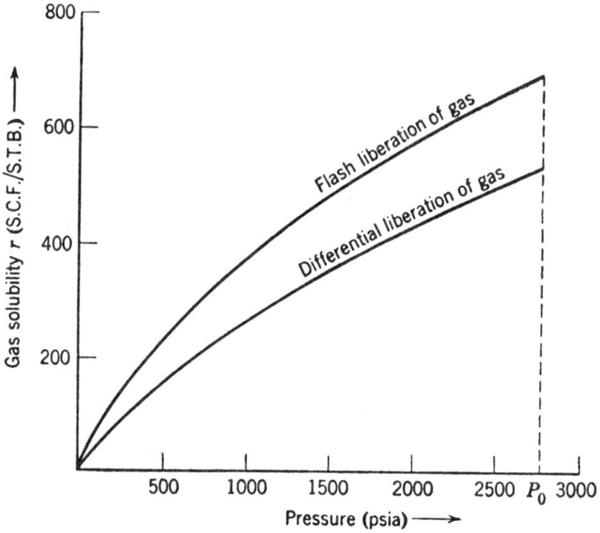

FIG. 56. Typical plot of r versus P showing differences obtained by flash and differential liberation of gas.

being higher at a given pressure. It is difficult to say which type of liberation is operative in a reservoir and in all probability both occur simultaneously.

It should be pointed out that the gas solubility could be calculated using the laws of solution behavior previously described in Chapter 5, provided sufficient data are available. It is necessary to have not only data for the overall composition of the system but complete and accurate equilibrium constant data as well. These data are seldom, if ever, available for a crude-oil system and values of gas solubility must be obtained either experimentally or by estimation. However, to illustrate the complexity of computations of this type the method is outlined below for a two-component system of known overall composition.

Step 1. Calculate the bubble-point pressure or saturation pressure using $\Sigma K_i x_i = 1$.

Step 2. Suppose r is to be calculated as a function of pressure when the gas is liberated by reducing the pressure in steps of 10 psia and the gas and oil separated after each step. Calculate the composition of the liquid and vapor at a pressure of $P_s - 10$ using

$$x_1 = \frac{1 - K_2}{K_1 - K_2}, \qquad x_2 = 1 - x_1, \qquad y_1 = K_1 x_1 \qquad \text{and} \qquad y_2 = 1 - y_1$$

Step 3. Calculate the amount of vapor at this pressure using

$$\frac{\text{Moles vapor}}{\text{Total moles}} = \frac{z - x}{y - x}$$

Step 4. Calculate the volume of vapor under standard conditions using

$$V = \text{moles vapor} \times 379$$

Step 5. Compute moles of liquid and repeat calculations at $P_s - 20$ for the residue liquid.

Step 6. Continue this process until atmospheric pressure is reached.

Step 7. At a given pressure add the volumes of vapor produced for each pressure reduction from the given pressure to atmospheric pressure. Calculate the volume of liquid remaining at atmospheric pressure and 60° F from the composition of the residue liquid and the liquid densities of each component. Calculate the volume of gas dissolved per barrel of stock tank oil.

For a multicomponent system this computation would be considerably more difficult since the composition and amounts of liquid and vapor at each pressure would have to be calculated by the trial-and-error method using equations 18 and 19 in Chapter 5.

Estimation of Solution Gas. An estimate of solution gas may be made with reasonable accuracy since considerable experimental data on the solubility of natural gas in crude oil are available and correlations between the solubility and the physical properties of the system have been made. A consideration of the variables which determine the solubility of a gas in a crude oil leads to the conclusion that the following are the most important. Furthermore, several generalizations regarding the effect of these variables on gas solubility can be made.

1. Effect of Pressure. Henry's Law predicts that at constant temperature the solubility of a gas in a given quantity of liquid is directly proportional to the pressure. Although the solubility of gas in a crude oil usually does not exhibit the linear dependence on pressure required by Henry's Law (see Figures 54 and 55), the solubility does increase with increasing pressure until the saturation pressure is reached.

2. Effect of Temperature. The solubility of natural gas in crude oil decreases with increasing temperature.

3. Effect of Gas Composition. The solubility of a gas in a crude oil decreases as the concentration of low molecular weight, highly volatile constituents in the gas increases. Since the gas gravity is determined by the molecular weight of the constituents a relationship between gas gravity and gas solubility should exist. It has been found that at a given temperature and pressure the solubility of a gas in a given crude oil decreases with decreasing specific gravity of the gas.

4. Effect of Oil Composition. Experimental data on the gas solubility in numerous crude oils indicate that the solubility increases as the specific gravity of the oil decreases. A low liquid gravity indicates the presence of appreciable concentrations of low molecular weight liquid hydrocarbons. Consequently there is a greater chemical similarity between the gas and the oil and a greater gas solubility is to be expected.

In the petroleum industry it is customary to express the specific gravity of crude oils in terms of API gravity. The relationship between API gravity and specific gravity measured at 60° F is expressed by the equation

$$^\circ\,\text{API gravity} = \frac{141.5}{\text{Specific gravity } 60^\circ\,\text{F}/60^\circ\,\text{F}} - 131.5$$

According to this equation the API gravity of water is 10 and the API gravity increases as the specific gravity decreases. In terms of the API gravity, it is apparent that the generalization regarding the effect of oil composition on gas solubility may be stated: At a given temperature and pressure the solubility of a gas in a crude oil increases with increasing API oil gravity.

These qualitative considerations concerning the effect of the above four variables on gas solubility can be correlated so that estimates of the solution gas can be made. The simplest correlation expresses the gas solubility as a function of pressure and the API gravity of the stock tank oil. This correlation is shown in Figure 57. The increased gas solubility with increasing pressure and API gravity is

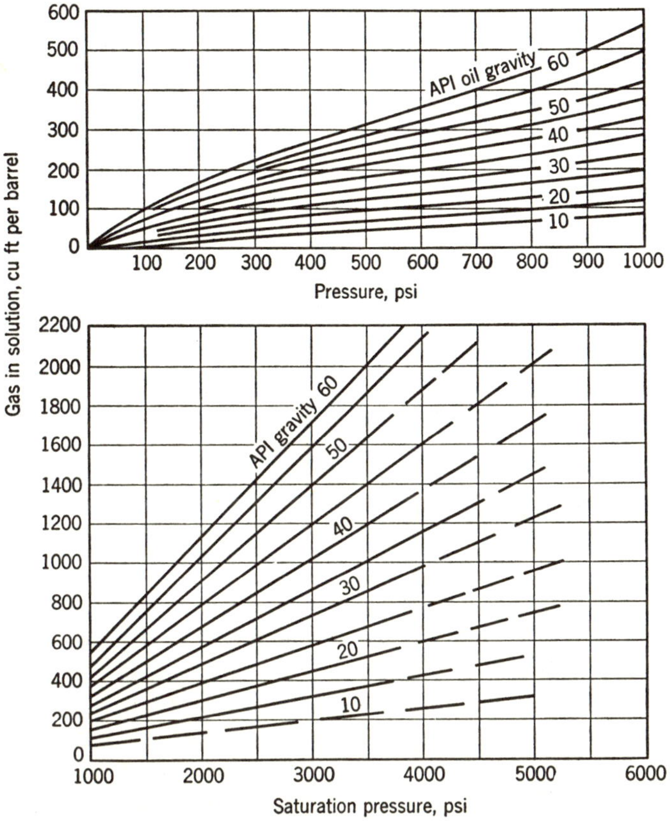

Fig. 57. Solution gas as a function of pressure and API oil gravity. (Beal, *Viscosity of Air, Water, Natural Gas, Crude Oil, and Its Associated Gases at Oil-Field Temperatures and Pressures, Trans. AIME, 165,* 1946, p. 106.)

clearly evident in this chart. It should be emphasized that since the effects of gas gravity and temperature are being neglected in this simple correlation errors higher than 25% may be involved.

Example. Estimate the solution gas in a 40 API oil at 2500 psia.

From Figure 57 the required solubility is found to be 850 S.C.F. per barrel of stock tank oil.

A more accurate method of predicting gas solubility is shown in Figure 58. In this correlation, which was originally proposed by

Standing, the solubility is expressed as a function of pressure, temperature, gas gravity, and oil gravity. The use of Figure 58 may best be illustrated by means of an example.

EXAMPLE. Using Standing's correlation estimate the solubility of 0.75 gravity gas in 30 API oil at a reservoir pressure of 1930 psia and a reservoir temperature of 200° F.

To use Figure 58 start at the right side of the chart and proceed vertically along the 1930 psia pressure line to a temperature of 200° F. From this point proceed horizontally to the left to the 30° API oil gravity line. Then continue vertically to the 0.75 gas gravity line. Finally, proceed horizontally to the left and the required gas solubility is found to be 350 S.C.F./stock tank barrel.

The Oil-Formation Volume Factor. If the pressure on a sample of reservoir oil is decreased below its saturation pressure gas is evolved

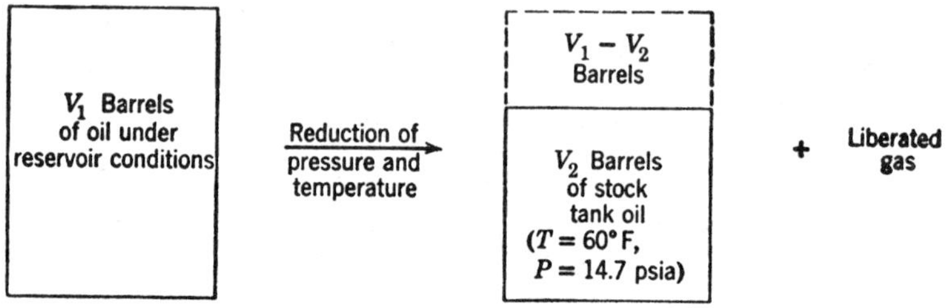

FIG. 59. Diagram showing relationship between oil volume under reservoir conditions and under stock tank conditions.

and the volume of the residual oil decreases. The shrinkage in oil volume upon liberation of gas may be expressed as a fractional volume change based either on the original or the final oil volume. This change in oil volume may also be expressed as a relative volume ratio. Here again the basis may be either the original or the final oil volume. Consider the process shown in Figure 59. A sample of reservoir liquid has a volume of V_1 barrels under reservoir conditions of temperature and pressure. When this liquid is brought to stock tank conditions the volume is reduced to V_2 barrels. The following four relationships may be used to express the change in volume.

$$\text{Shrinkage based on final oil volume} = Sh_2 = \frac{V_1 - V_2}{V_2} \tag{2}$$

$$\text{Shrinkage based on original oil volume} = Sh_1 = \frac{V_1 - V_2}{V_1} \tag{3}$$

$$\text{Formation volume factor} = \beta = \frac{V_1}{V_2} \tag{4}$$

$$\text{Shrinkage factor} = \gamma = \frac{V_2}{V_1} \tag{5}$$

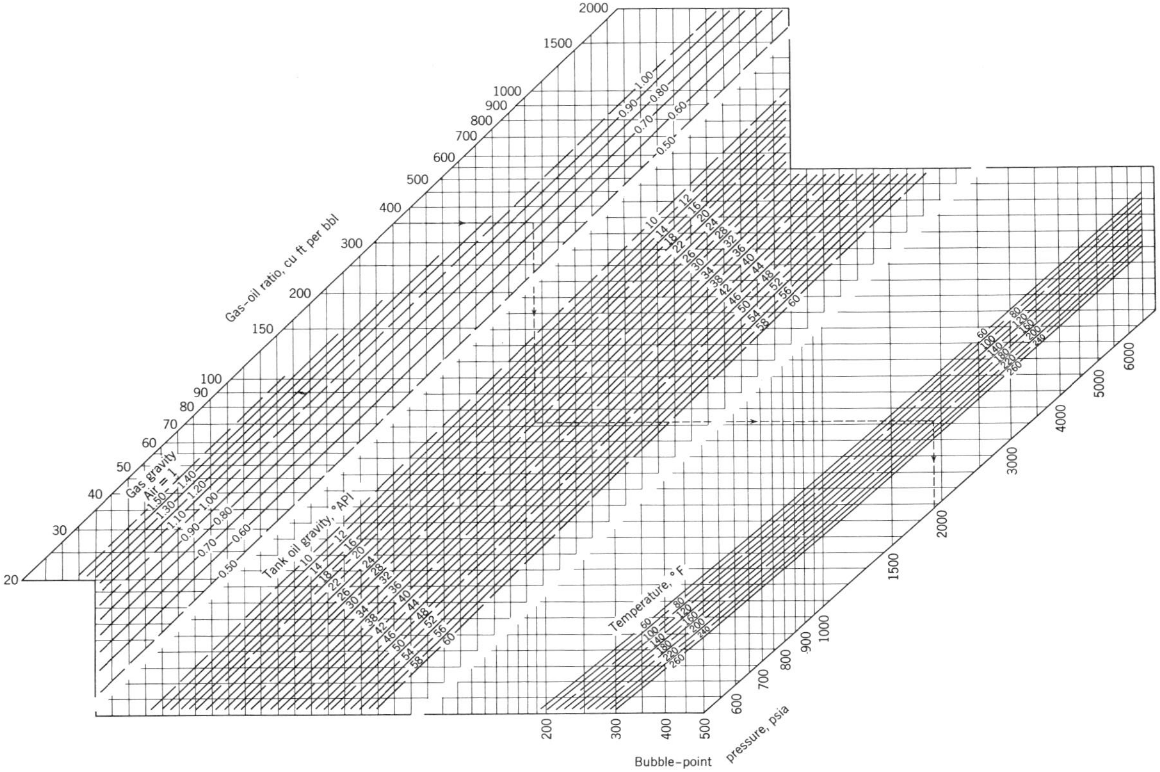

Fig. 58. Gas solubility as a function of pressure, temperature, gas gravity, and oil gravity. (Copyright 1947, California Research Corporation, reprinted by permission.)

It is obvious that the following relationships exist between these quantities

$$\beta = 1 + Sh_2 = \frac{1}{\gamma} = \frac{1}{1 - Sh_1}$$

and

$$\gamma = 1 - Sh_1$$

In engineering calculations the formation volume factor β is most commonly used to express the change in liquid volume with pressure. β is defined by equation 4 or by an equivalent definition as the volume

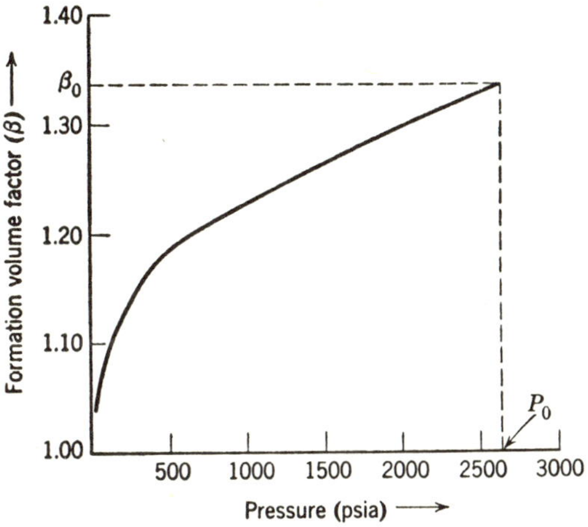

FIG. 60. Typical plot showing the dependence of the formation volume factor on pressure for a saturated crude oil.

in barrels at reservoir pressure and temperature occupied by one barrel of stock tank oil plus the gas in solution at that temperature and pressure. The change in β with pressure for a typical saturated crude oil is shown in Figure 60. The original reservoir pressure is P_0 and original value of the formation volume factor is represented by β_0. As the pressure is reduced below P_0 solution gas is evolved, the volume of the oil is reduced, and the value of β is decreased. When the pressure is reduced to atmospheric the value of β is nearly equal to one. ($\beta = 1$ when the reservoir temperature is 60° F.) For a typical undersaturated crude the value of β as a function of pressure is given by a curve similar to that shown in Figure 61. The original reservoir pressure is P_0 and the saturation or bubble-point pressure is P_s. As the pressure is reduced below P_0 the volume of the all-liquid system increases due to expansion. This behavior is reflected by an increase in the value of β. The value of β continues to increase until the saturation pressure is reached. At P_s the value of β is a maximum represented by β_s A further reduction in pressure below P_s results in the evolution of gas and the value of β decreases. The behavior below P_s is similar to that for the saturated crude oil shown in Figure 60.

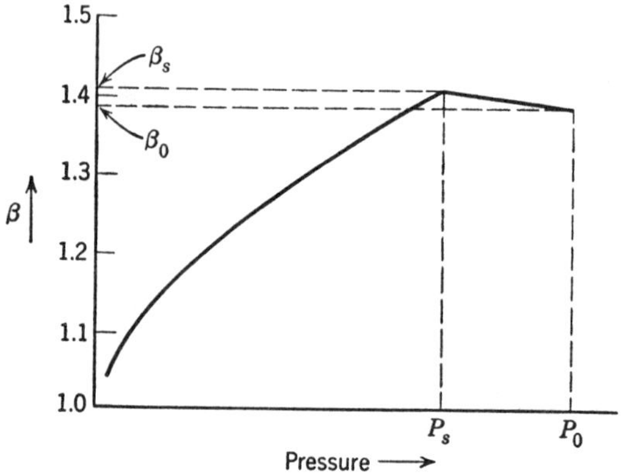

FIG. 61. Illustrative plot of β versus pressure for an undersaturated crude oil.

As was pointed out in the discussion of solution gas the value of r depends on the method of gas liberation. It is evident that the value of the formation volume factor is also dependent on the method of gas liberation, as is shown in Figure 62. Since less gas is evolved on differential liberation the residual oil volume is greater. Consequently, the differential formation volume factor is less than the flash formation volume factor as shown.

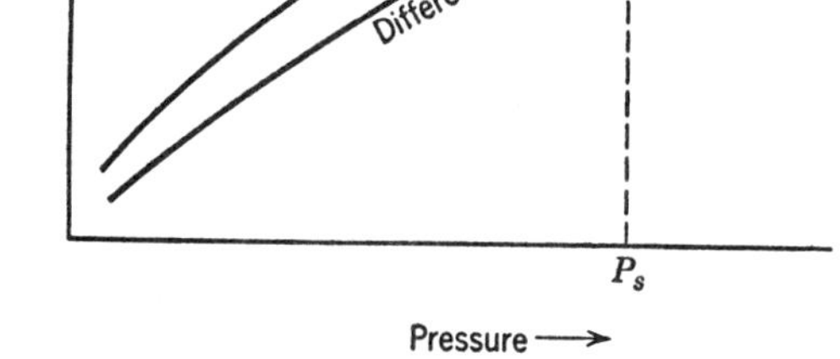

FIG. 62. Illustrative plot showing dependence of the formation volume factor on the method of gas liberation.

If the composition of the crude oil and the equilibrium constants of the components were known β could be calculated using the principles outlined in the previous chapter. The composition and amount of liquid and vapor could be calculated at any temperature and pressure. Using these results, the volume of the liquid could be calculated from the known densities of the various components in the liquid. Similarly, the volume of the residue liquid at stock tank conditions could be computed. β would be the ratio of these two volumes. Again, for most crude oils, calculations of this type are impossible since the composition of the original crude oil is seldom known. Consequently, it is usually necessary to make experimental determinations of β versus P in the laboratory using a sample of the reservoir crude oil. If experimental data are not available it is possible to estimate a value of β using one of the following methods for this purpose.

Estimation of the Formation Volume Factor. *Method 1.* The simplest method of estimating the formation volume factor requires

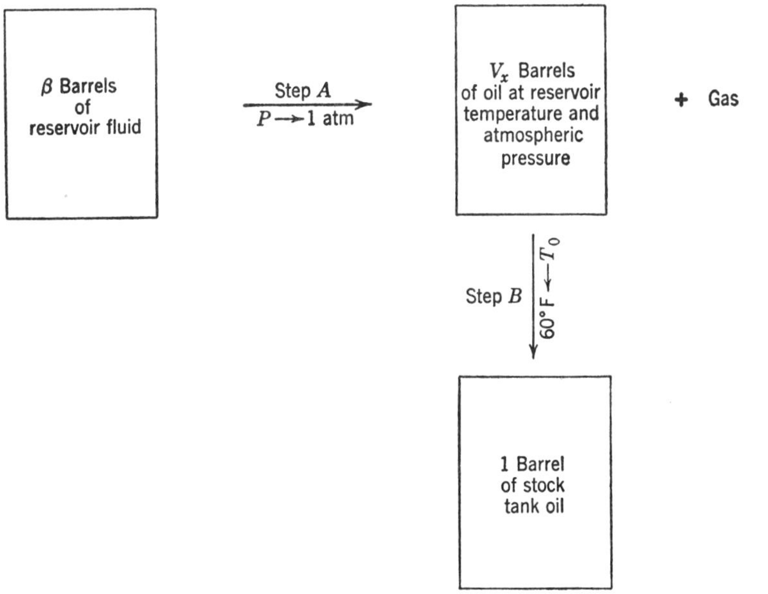

FIG. 63. Stepwise change in oil volume.

a knowledge of the API gravity of the stock tank oil, the reservoir temperature, and the pressure at which the formation volume factor is desired. For this correlation assume that the change in oil volume from reservoir conditions to stock tank conditions takes place in two steps. In step A the pressure is reduced from reservoir pressure to atmospheric at constant temperature with the evolution of solution gas. In step B the temperature is reduced from reservoir temperature to 60° F at a constant pressure of one atmosphere. This process is illustrated in Figure 63 for β barrels of reservoir oil. For step B the shrinkage based on the final volume is

$$Sh_B = \frac{V_x - 1}{1} \quad \text{or} \quad V_x = 1 + Sh_B \tag{6}$$

Similarly for step A the shrinkage based on the final volume is

$$Sh_A = \frac{\beta - V_x}{V_x} \tag{7}$$

Eliminating the intermediate volume V_x from equations 6 and 7 and solving for β one obtains

$$\beta = (1 + Sh_A)(1 + Sh_B) \tag{8}$$

Consequently, if Sh_A and Sh_B are known β may be calculated. The value of Sh_A is determined largely by the quantity of gas evolved. A plot of Sh_A versus gas solubility is given in Figure 64. Values of Sh_B

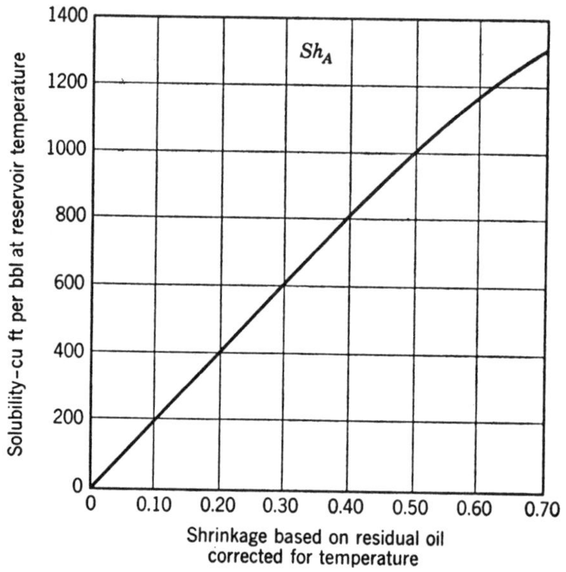

FIG. 64. Sh_A as a function of gas solubility. (Katz, *API Drilling and Production Practice,* 1942, p. 144.)

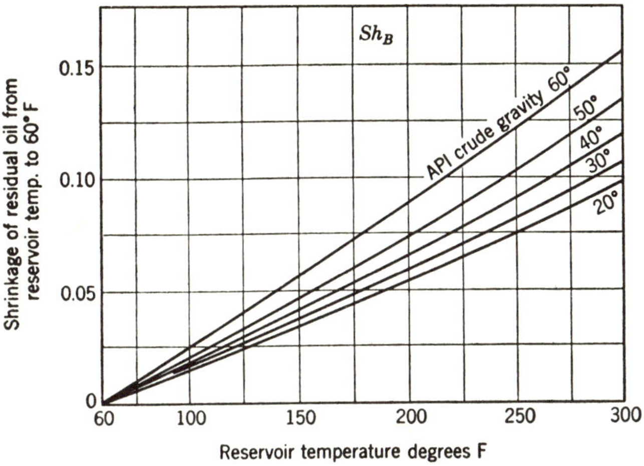

FIG. 65. Sh_B as a function of reservoir temperature. (Katz, *API Drilling and Production Practice,* 1942, p. 144.)

as a function of reservoir temperature for various API gravity crudes are given in Figure 65. An example illustrating the use of Figures 64 and 65 to estimate β is given below.

EXAMPLE. Estimate β for a 30° API oil at a reservoir temperature of 150° F and a pressure of 2500 psia.

From Figure 57, the solution gas for a 30° API oil at a pressure of 2500 psia is estimated to be 600 S.C.F. per stock tank barrel of oil. Consequently, Sh_A is 0.29 from Figure 64. At a reservoir temperature of 150° F, Sh_B is 0.036 for a 30° API oil as shown in Figure 65. Therefore, according to equation 8, $\beta = (1 + 0.29)(1 + 0.036) = 1.34$

This method of estimating the formation volume factor is accurate within about 15%.

Method 2. If the gas gravity, oil gravity, gas solubility, and reservoir temperature and pressure are known the formation volume factor may be estimated with a probable error of about 5% using the following method. This method of correlation is based on the premise that a gas dissolved in a crude oil exhibits an *apparent* liquid density which is a function of the gas gravity and the API gravity of the oil. The apparent density of the dissolved gas at standard conditions of temperature and pressure is estimated from the empirical correlation curves shown in Figure 66. From the known gas solubility and gas gravity the weight of gas dissolved in one barrel of stock tank oil may be calculated. From this weight and the apparent gas density estimated from Figure 66 the increase in volume of the stock tank oil may be computed. Furthermore, since the total weight of the oil and gas

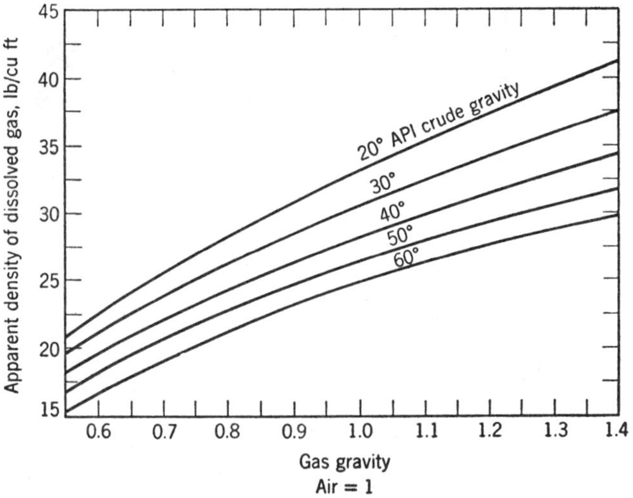

FIG. 66. Apparent liquid density for various gases in various API gravity oils. (Katz, *API Drilling and Production Practice,* 1942, p. 140.)

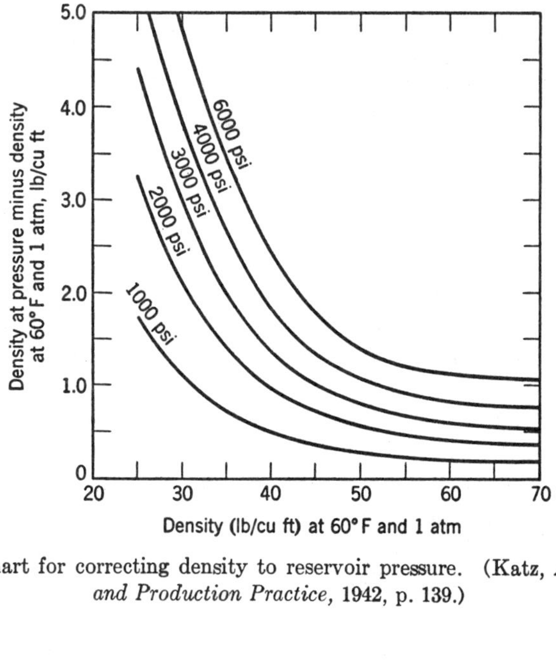

FIG. 67. Chart for correcting density to reservoir pressure. (Katz, *API Drilling and Production Practice,* 1942, p. 139.)

may be calculated, this weight divided by the combined volume of the stock tank oil and gas gives the density of the system at standard conditions. This density is corrected to reservoir conditions with the aid of Figures 67 and 68. The total weight of oil and gas divided by the density obtained in this manner represents cubic feet of reservoir volume occupied by one stock tank barrel of oil. Consequently this

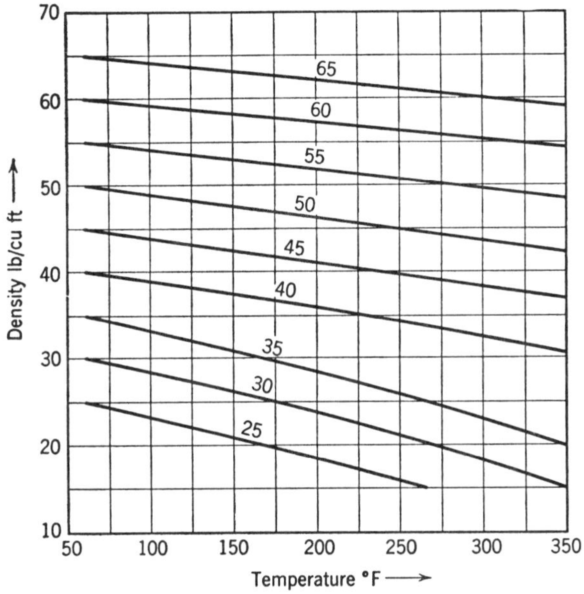

FIG. 68. Density change of crude oils with temperature. Figures on curves represent oil density in lb/cu ft at 60° F. (After Katz, *API Drilling and Production Practice*, 1942, p. 139.)

volume divided by 5.62 is equal to β. The following example illustrates this method of estimating the formation volume factor by Method 2.

EXAMPLE. Estimate the formation volume factor at 1000 psia and 150° F for a 30° API gravity stock tank oil. The gas solubility at this pressure and temperature is known to be 225 S.C.F. of 0.65 gravity gas per barrel of stock tank oil.

The molecular weight of this gas is

$$29 \times 0.65 = 18.85$$

Since one pound-mole of gas occupies a volume of 379 standard cubic feet the number of moles of gas dissolved per stock tank barrel is

$$\frac{225}{379} = 0.594 \text{ pound-moles}$$

From Figure 66 the apparent density of a 0.65 gravity gas in 30° API gravity oil is 22.5 lb per cu ft. Consequently, the increase in oil volume due to solution gas is given by

$$\frac{18.85 \times 0.594}{22.5} = 0.497 \text{ cu ft/bbl}$$

The density of the stock tank oil plus solution gas at standard conditions is equal to the weight of the oil and the gas divided by the combined volume. Since a 30° API oil gravity is equivalent to a specific gravity of 0.876 the density of the combined oil-gas system at standard conditions is given by

$$\frac{0.876 \times 62.5 \times 5.62 + 18.85 \times 0.594}{5.62 + 0.497} = \frac{318.7}{6.12} = 52.2 \text{ lb/cu ft}$$

When this density is corrected to reservoir pressure and temperature using Figures 67 and 68 a value of 50.5 lb/cu ft is obtained. In making this correction it must be remembered that an increase in pressure causes an increase in density but an increase in temperature causes a decrease in density. The volume of reservoir oil is given by

$$\frac{318.7}{50.5} = 6.32 \text{ cu ft per barrel of stock tank oil}$$

and β is

$$\frac{6.32}{5.62} = 1.12$$

Method 3. If the stock tank oil gravity, the gas solubility, and the gas *composition* are known, the formation volume factor may be estimated at any given reservoir temperature and pressure with an accuracy which approaches that of experimental determinations. This method is based on the assumption that, when a hydrocarbon gas dissolves in a crude oil, the combined volume of the system is equal to the volume of the oil plus that of the gas if it existed in the liquid state. From the known solubility and composition of the gas the weight and liquid volume of each constituent may be computed, the latter being calculated from the liquid density of each constituent at 60° F and at a pressure equal to its vapor pressure. The weight of the combined system divided by the combined liquid volume represents the density at 60° F and one atmosphere pressure. This density is corrected to reservoir temperature and pressure using Figures 67 and 68, and the value of β is computed in the same manner as described in the preceding method for estimating the formation volume factor.

For methane and ethane it was found that an *apparent* liquid density must be employed to calculate their liquid volumes. Furthermore, these apparent densities are a function of the concentrations of methane and ethane in the liquid and the combined density of the constituents other than methane and ethane. The apparent densities of methane and ethane as a function of their weight per cents in the liquid and the density of the remainder of the system is given in Figure 69. To use this chart for a system containing both methane

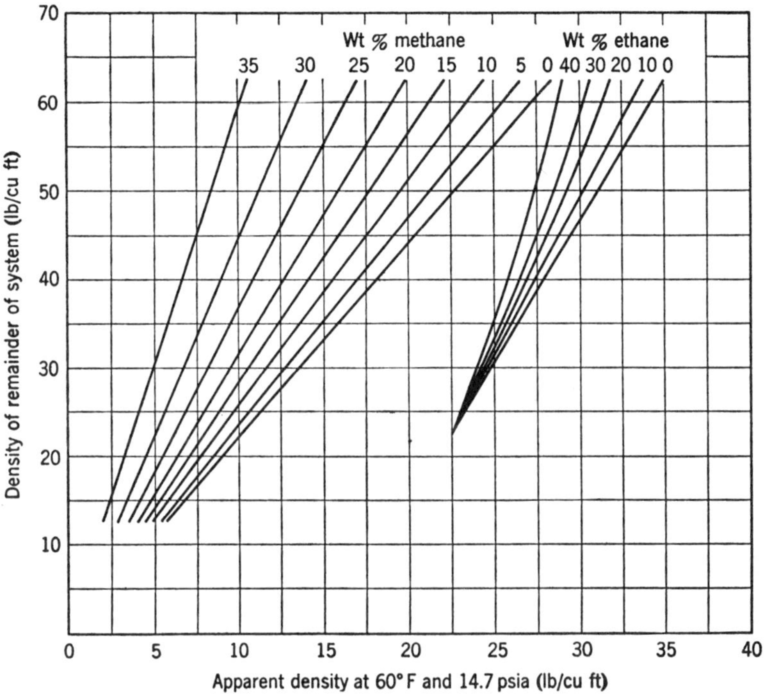

FIG. 69. Apparent density of methane and ethane. (After Katz, *API Drilling and Production Practice*, 1942, p. 139.)

and ethane it is necessary to compute first the density of the system without these two components and then find the apparent density of ethane from the chart. The density of the system without methane is then calculated and this density is used to find the apparent density of methane.

The principal difference between this method and Method 2 for estimating β is that in Method 2 the apparent liquid density of the gas as a *whole* is estimated from its gravity and the gravity of the oil in which it is dissolved. In Method 3 the volume of each gas in solution is calculated. An example computation of β using Method 3 is given in the following.

EXAMPLE. A natural gas has the following composition:

Component	Volume %
Methane	75
Ethane	10
Propane	7
Butane	6
Pentane	1
Hexane	1

The solubility of this gas in a 40° API crude is 800 S.C.F. per stock tank barrel at 2500 psia and 150° F. Estimate β.

The following calculations are made as indicated:

(1)	(2)	(3)	(4)	(5)
Component	S.C.F. dissolved per bbl (800 × volume fraction)	lb dissolved per bbl [(S.C.F./379) × MW]	Density (lb/cu ft) of liquid at 60° F and vapor pressure (see Appendix *B*)	Volume (cu ft) in liquid state $\left(\frac{\text{column 3}}{\text{column 4}}\right)$
Methane	600	25.3	20.6 *	1.230
Ethane	80	6.3	31.2 *	0.202
Propane	56	6.5	31.6	0.206
Butane	48	7.3	36.4	0.200
Pentane	8	1.5	39.3	0.038
Hexane	8	1.8	41.3	0.044

Total weight of gas = 48.7 lb

Weight of gas without methane = 23.4 lb
Weight of gas without methane and ethane = 17.1 lb

* These values are *apparent* densities and are calculated as shown below.

Volume of crude oil = 1 barrel = 5.62 cu ft
Weight of crude oil = 350 × 0.825 = 289 lb
Density of crude oil and gas system without methane and ethane =

$$\frac{289 + 17.1}{5.62 + 0.488} = 50.1 \text{ lb/cu ft}$$

$$\text{Weight \% ethane} = \frac{6.3}{289 + 23.4} \times 100 = 2.0\%$$

Apparent density of ethane (from Figure 69) = 31.2 lb/cu ft

$$\text{Density of crude oil and gas system without methane} = \frac{289 + 23.4}{5.62 + 0.488 + 0.202}$$
$$= 49.4 \text{ lb/cu ft}$$

$$\text{Weight \% methane} = \frac{25.3}{289 + 48.7} \times 100 = 7.5\%$$

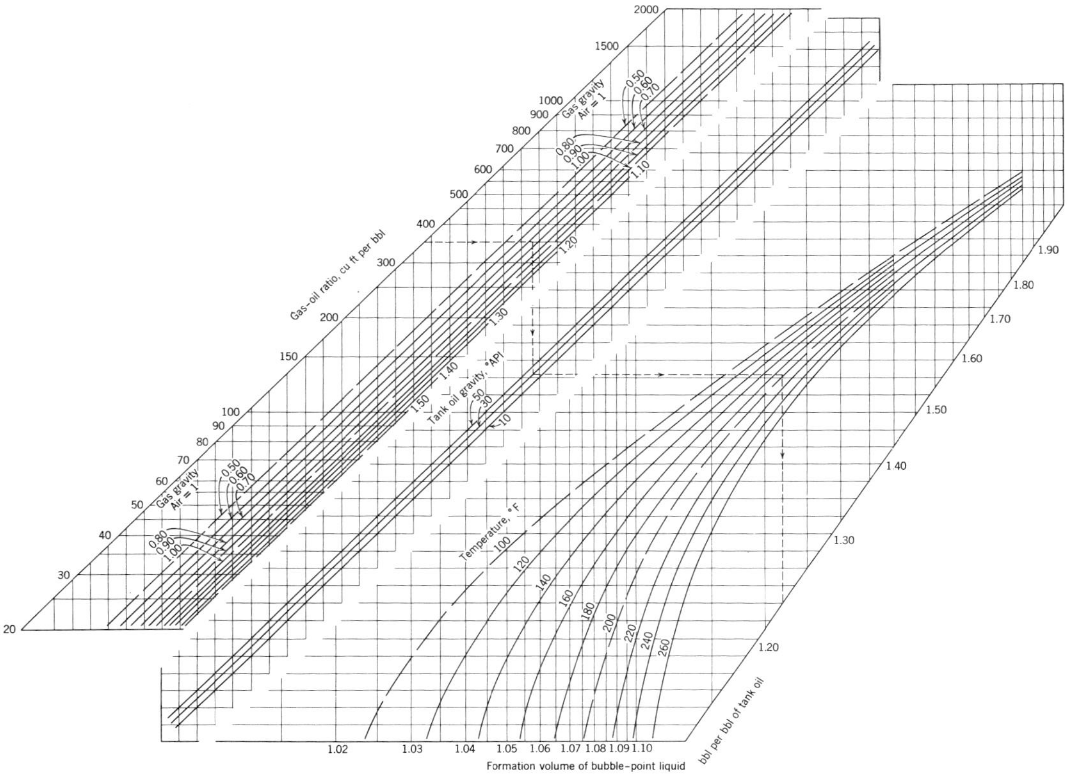

Fig. 70. Oil-formation volume factor as a function of gas solubility, temperature, gas gravity, and oil gravity. (Copyright 1947, California Research Corporation, reprinted by permission.)

Apparent density of methane (from Figure 69) = 20.6 lb/cu ft

$$\text{Density of reservoir fluid at 60° F and 14.7 psia} = \frac{289 + 48.7}{5.62 + 0.488 + 0.202 + 1.23}$$

$$= \frac{337.7}{7.54} = 44.8 \text{ lb/cu ft}$$

When this density is corrected to 2500 psia and 150° F using Figures 67 and 68 the density under reservoir conditions is found to be 43.1 lb/cu ft. This gives a value of β equal to

$$\frac{337.7}{43.1 \times 5.62} = 1.395$$

Method 4. The three methods for estimating the formation volume factor that have been presented were first proposed by Katz. It is readily apparent that the computation involved becomes increasingly complex as the data available allows an increasingly accurate estimate to be made. A fourth method, which is relatively simple to apply, has been developed by Standing and requires a knowledge of the gas solubility, the gas and oil gravity, and the reservoir temperature. This method is based on experimental data obtained from 22 different California crude-oil–natural-gas mixtures and for the data employed the average error between the experimental values and those obtained by estimation was about 1.2%.

The chart shown in Figure 70 is used to estimate β. The use of Figure 70 may best be illustrated by means of an example.

EXAMPLE. Estimate the oil formation volume factor at 200° F for a 30° API oil which contains 350 S.C.F. of 0.75 gravity gas in solution per barrel of stock tank oil.

Enter the chart in Figure 70 from the left at 350 cubic feet per barrel and proceed horizontally to the 0.75 gas gravity line. Drop vertically to the 30° API oil gravity line and then proceed horizontally to the 200° F reservoir temperature line. By dropping vertically again the required oil formation volume factor is found to be 1.22.

The four methods for estimating the oil-formation volume factor that have been described are not applicable to systems above the bubble point. As already pointed out, the decrease in β with pressure above the saturation pressure is a result of the compression of the all-liquid system. If the coefficient of compression of this liquid is known it is possible to compute the values of β above the saturation pressure in the following manner.

By definition the average coefficient of compression is equal to the change in volume with pressure per unit volume. Stated mathematically

$$C = -\frac{V_2 - V_1}{V_1(P_2 - P_1)}$$

where C is the average coefficient of compression over the pressure range $P_2 - P_1$, V_2 is the volume at P_2 and V_1 is the volume at P_1. The negative sign preceding the right-hand side of this equation is necessary if the coefficient of compression is to be designated as a positive number. This equation may be written in the form

$$V_2 = V_1[1 - C(P_2 - P_1)]$$

Dividing both sides of this equation by the stock tank volume gives

$$\beta_2 = \beta_1[1 - C(P_2 - P_1)]$$

If the saturation pressure is chosen as the initial pressure this equation becomes

$$\beta_2 = \beta_s[1 - C(P_2 - P_s)] \tag{9}$$

Obviously, this equation can be used to calculate the formation volume factor at any pressure P_2 above the saturation pressure provided C

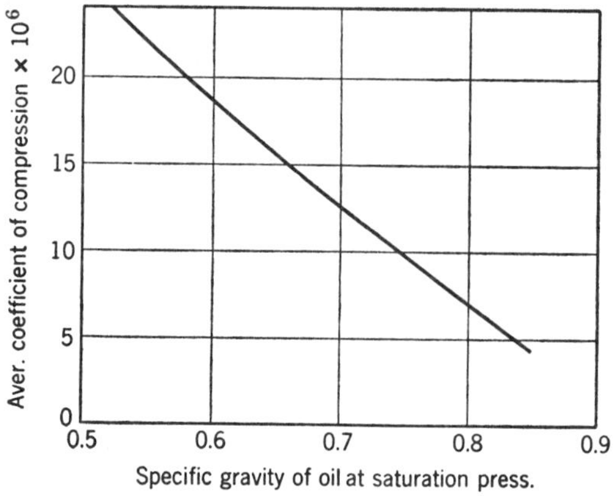

FIG. 71. Average coefficient of compression as a function of oil gravity at the saturation pressure. (From Calhoun, John C., Jr., *Fundamentals of Reservoir Engineering,* Copyright 1953, University of Oklahoma Press, p. 35.)

and P_s and β_s are known. Values of C computed from data available in the literature show a trend of C versus specific gravity of the reservoir oil *at the bubble point,* as shown in Figure 71. Consequently if this specific gravity is known values of β above the bubble point may be computed as shown in the example below.

EXAMPLE. An oil has a saturation pressure of 2000 psia and a specific gravity of 0.75 at the saturation pressure. If β_s is 1.46 estimate β at 3500 psia.

From Figure 71 the coefficient of compressibility of a 0.75 gravity oil is 1×10^{-5}. Consequently, according to equation 9

$$\beta = 1.46[1 - 10^{-5}(3500 - 2000)] = 1.44$$

In order to employ Figure 71 effectively it is necessary to know the specific gravity of the reservoir oil at the saturation pressure. If β_s, r_0, the gas gravity, and the stock tank oil gravity are known, the specific gravity of the reservoir oil at P_s may be calculated in the following manner. The weight of gas dissolved in one barrel of stock tank oil is given by

$$\frac{r_0}{379} \times \text{gas gravity} \times 29$$

This weight plus the weight of one stock tank barrel of oil represents the weight of β_s barrels of reservoir fluid. Since the weight of an equal volume of water is $350\beta_s$, the specific gravity of the oil under reservoir conditions is computed by dividing the weight of β_s barrels of reservoir fluid by the weight of β_s barrels of water.

EXAMPLE. At the saturation pressure one stock tank bbl of 30° API oil dissolves 775 S.C.F. of 0.75 gravity gas. At the saturation pressure the formation volume factor is 1.42. Calculate the specific gravity of the oil at the saturation pressure.

Weight of gas dissolved $= \frac{775}{379} \times 0.75 \times 29 = 44.5$ lb
Weight of one stock tank barrel of 30° API oil $= 350 \times 0.876 = 307$ lb
Weight of β_s barrels of reservoir fluid $= 307 + 44.5 = 351.5$ lb
Weight of β_s barrels of water $= 1.42 \times 350 = 497$ lb
Specific gravity of reservoir oil at the saturation pressure $= 351.5/497 = 0.706$

The Two-Phase Formation Volume Factor (u). In reservoir engineering calculations it is sometimes convenient to know the volume occupied in the reservoir by one stock tank barrel of oil plus the free gas that was originally dissolved in it. This volume is known as the two-phase formation volume factor and is given the symbol u. It is apparent that the value of u is determined by the values of the reservoir fluid characteristics previously described. Expressed mathematically u is defined by the following equation

$$u = \beta + (r_0 - r)v$$

since the formation volume factor β represents the liquid volume of one stock tank barrel at reservoir conditions and $(r_0 - r)v$ denotes the volume of the free gas under reservoir conditions that was originally in solution. The latter follows from the fact that $(r_0 - r)$ represents the number of standard cubic feet of gas that have come out of solution and v is the previously described conversion factor for computing barrels of free gas space occupied by one standard cubic foot of gas in the reservoir.

The meaning of the two-phase formation volume factor is shown diagrammatically in Figure 72 which illustrates the volume changes

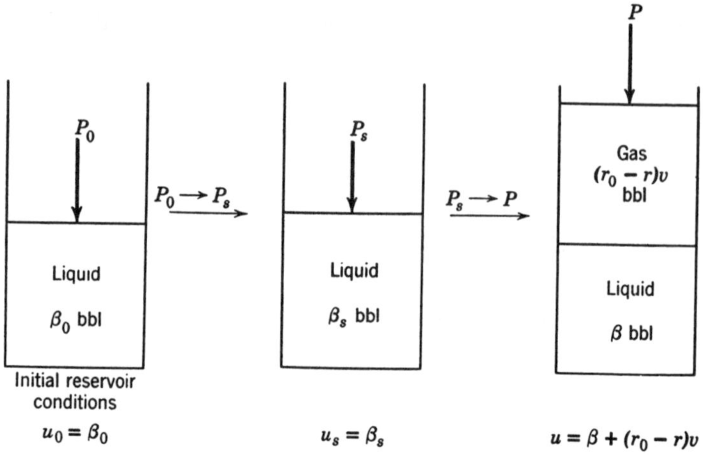

FIG. 72. Volume changes which occur when pressure is decreased on β_0 barrels of reservoir fluid.

that occur when pressure is released from a quantity of an undersaturated reservoir oil sufficient to yield one barrel of stock tank oil.

A typical plot of u as a function of pressure for an undersaturated crude is shown in Figure 73. The portion of the curve between P_0 and

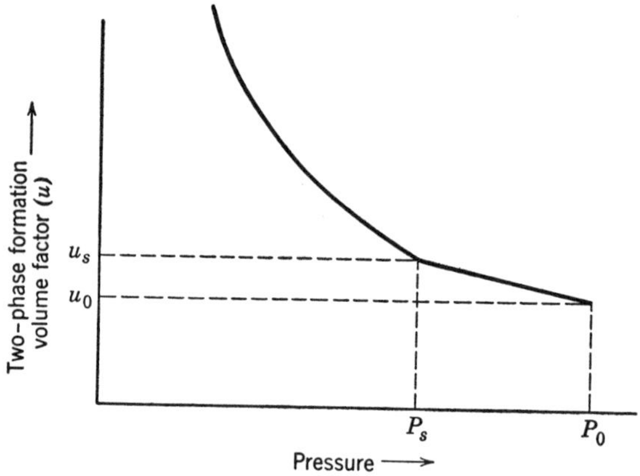

FIG. 73. Illustrative plot of u as a function of pressure for an undersaturated crude.

P_s is coincident with the corresponding β versus pressure curve since the system is all liquid in this pressure range. As the pressure is reduced below P_s the value of u increases due to the evolution of solution gas. In the special case of a reservoir whose temperature is 60° F u reaches a value of $1 + (r_0/5.62)$ at 14.7 psia. It should also be noted that u continually increases with decreasing pressure. A plot of u as a function of pressure for a saturated crude has the characteristics shown in Figure 74. The β versus pressure curve is also included in this figure for the sake of comparison.

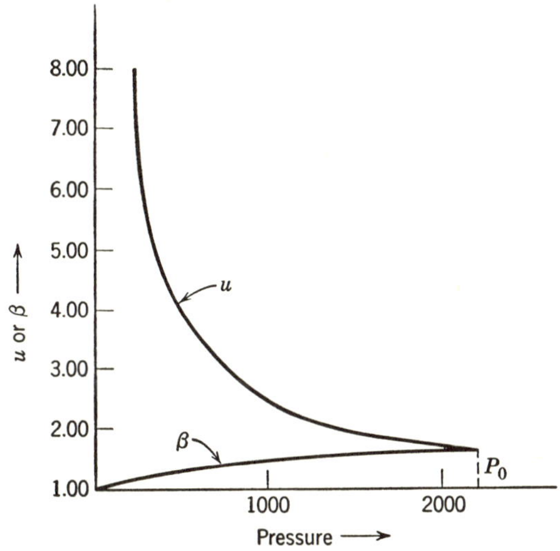

Fig. 74. Illustrative plot of u and β as a function of pressure.

The concept of the two-phase formation volume factor is introduced mainly for convenience. As will be shown in the next chapter, many of the fundamental reservoir equations are simplified if they are expressed in terms of u.

Example. An all-liquid sample of reservoir fluid occupied a volume of 331 cc at an original reservoir temperature and pressure of 150° F and 2500 psia. When the pressure was reduced to 2000 psia gas was evolved and the total volume of the gas and liquid was 351 cc at 150° F. The gas was bled off at constant pressure and temperature and the residue liquid occupied a volume of 306 cc. The pressure and temperature were reduced to 14.7 psia and 60° F and 225 cc of stock tank oil were obtained. Calculate β_0, β at 2000 psia and 150° F, u_0 and u at 2000 psia and 150° F.

From the definitions of β and u the following results are obtained.

$$\beta_0 = \tfrac{331}{225} = 1.47 = u_0$$

$$u\ (2000 \text{ psia and } 150°\text{ F}) = \tfrac{351}{225} = 1.56$$

$$\beta\ (2000 \text{ psia and } 150°\text{ F}) = \tfrac{306}{225} = 1.36$$

Viscosities of Reservoir Fluids (μ_o, μ_g). The viscosity of a fluid, which is a measure of its resistance to flow, is defined as the force in dynes on unit area of either of two horizontal planes unit distance apart, one of which is fixed while the other moves with unit velocity, the space between the planes being filled with the viscous fluid.

It will be necessary to discuss the effects of pressure and temperature on the viscosity of liquids and gases. In the case of hydrocarbon *liquids* certain generalizations can be made: (1) Viscosity decreases

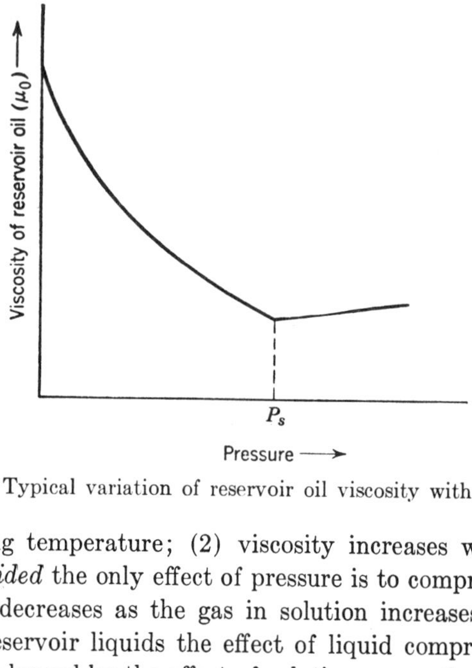

FIG. 75. Typical variation of reservoir oil viscosity with pressure.

with increasing temperature; (2) viscosity increases with increasing pressure, *provided* the only effect of pressure is to compress the liquid; (3) viscosity decreases as the gas in solution increases.

For most reservoir liquids the effect of liquid compression is more than counterbalanced by the effect of solution gas so that the viscosity decreases with pressure until the saturation pressure is reached. A further increase in pressure will cause an increase in viscosity due to compression of the liquid as is shown in Figure 75.

If experimental data for the crude oil viscosity are not available an estimate can be made in the following manner. If the liquid is below the saturation pressure the viscosity is estimated with the aid of Figures 76 and 77. Figure 76 shows the viscosities at one atmosphere pressure for various API stock tank oils at various reservoir temperatures. Since crude oils vary widely in composition the prediction of viscosities by this simple correlation shows an average deviation of about 25% when compared to experimental values. After the viscosity is estimated at reservoir temperature the effect of solution gas on viscosity is estimated with the aid of Figure 77.

EXAMPLE. A 40 API gravity oil has 680 S.C.F. of gas in solution per barrel of stock tank oil at 2000 psia. Estimate the liquid viscosity if the reservoir temperature is 150°F.

From Figure 76 the viscosity at atmospheric pressure and 150° F is 1.6 c.p. for a 40 API stock tank oil. From Figure 77 it is seen that this oil would have a viscosity of 0.55 c.p. when 680 S.C.F. of gas are in solution.

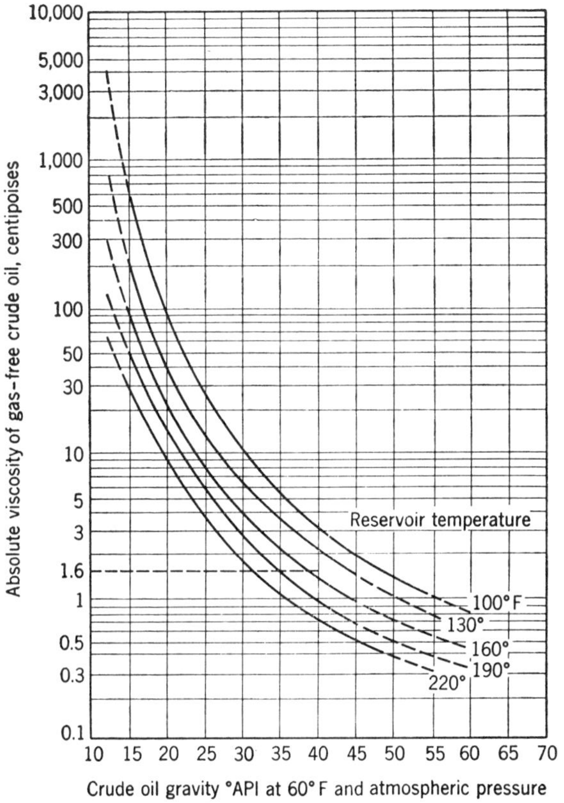

FIG. 76. Viscosity of reservoir oils at one atmosphere pressure. (After Beal, *Trans. AIME, 165,* 1946, p. 103.)

If the pressure is above the saturation pressure, the viscosity can be estimated with the aid of Figure 78. This chart is based on the viscosity at the saturation pressure which, if not known, may be estimated in the manner previously described. This chart clearly shows the increase in liquid viscosity due to compression of the liquid at pressures greater than the saturation pressure. It has been found that viscosity estimations above the saturation pressure are quite accurate, provided the viscosity at the saturation pressure is accurately known. Viscosities predicted from Figure 78 show an average deviation of less than 3% when compared with actual experimental values.

EXAMPLE. Estimate the viscosity at 4000 psia of an oil whose viscosity is 45 c.p. at a saturation pressure of 1800 psia.

From Figure 78 the viscosity of an oil whose viscosity is 45 c.p. at P_s is found to be 69 c.p. at a pressure 2200 psi above the bubble point.

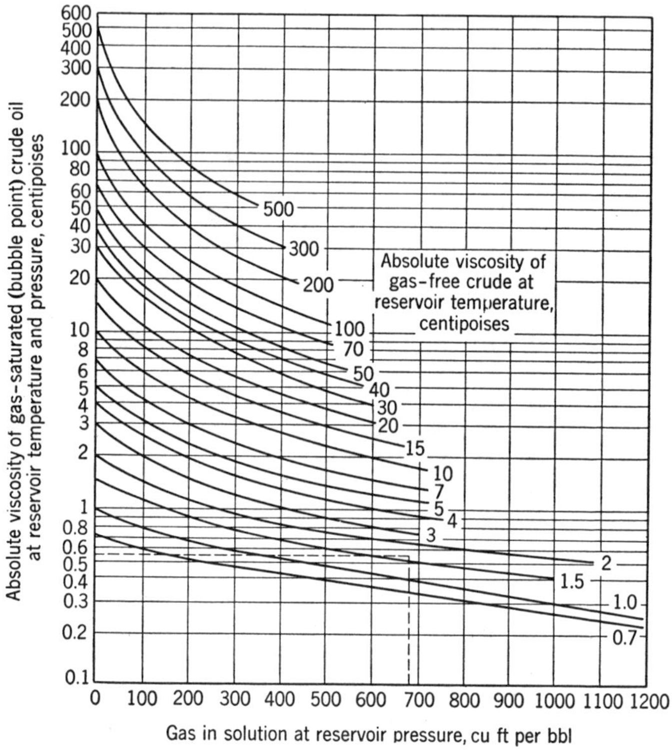

FIG. 77. Effect of solution gas on reservoir oil viscosity. (After Beal, *Trans. AIME, 165,* 1946, p. 105.)

The viscosities of gases are, in general, considerably lower than those of liquids. Certain generalizations can be made for the effect of pressure and temperature on the viscosity of a gas. For a perfect gas the viscosity *increases* with temperature. In this respect the behavior of a gas is the reverse of that of a liquid. Furthermore, the viscosity of a perfect gas is independent of pressure. This rather unexpected behavior of gases can be explained on the basis of the Kinetic

Theory of matter but this subject is beyond the scope of this book. It is interesting to note that these effects of temperature and pressure on the viscosity of gases were first predicted by theory and then experimentally verified. As the pressure on a perfect gas is increased it becomes imperfect and its behavior approaches that of a liquid. Con-

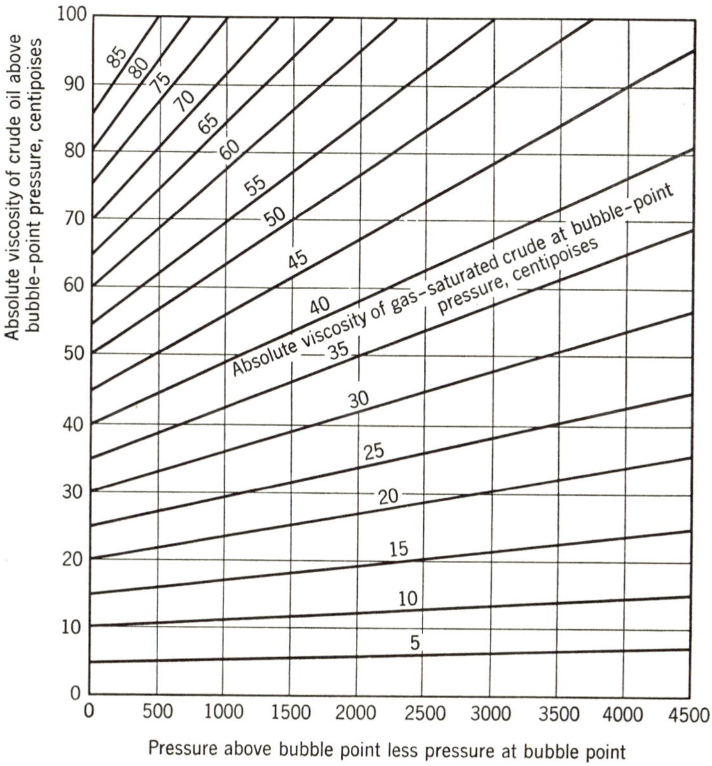

FIG. 78. Chart for estimating oil viscosity above the saturation pressure. (Beal, *Trans. AIME, 165,* 1946, p. 110.)

sequently, at the high pressures usually encountered in petroleum reservoirs the viscosity of gases may increase with pressure and decrease with temperature, as is the case with liquids.

The viscosity of various specific gravity gases as a function of temperature and pressure are shown in Figure 79. These charts enable a reasonably accurate estimate to be made if experimental data on the gas viscosity are not available. These charts were prepared from gas viscosity data obtained by Bicher and Katz on methane, propane, and methane-propane mixtures. A comparison between viscosities predicted with the aid of these charts and actual experimental values for several natural gases show an average deviation of about 6%.

EXAMPLE. Estimate the viscosity of a 0.75 gravity gas at 3000 psia and a temperature of 100° F.

From Figure 79 (parts 2 and 3) the required viscosity is found to be 0.0255 c.p. by interpolation.

If the composition of the gas is known, a more accurate estimate of the gas viscosity can be made by a method proposed by Carr, Kobayashi, and Burrows. The correlating procedure is based on the Law of Corresponding States. Consider the ratio μ/μ_1 where μ is the viscosity of the gas at the reduced reservoir temperature and pressure

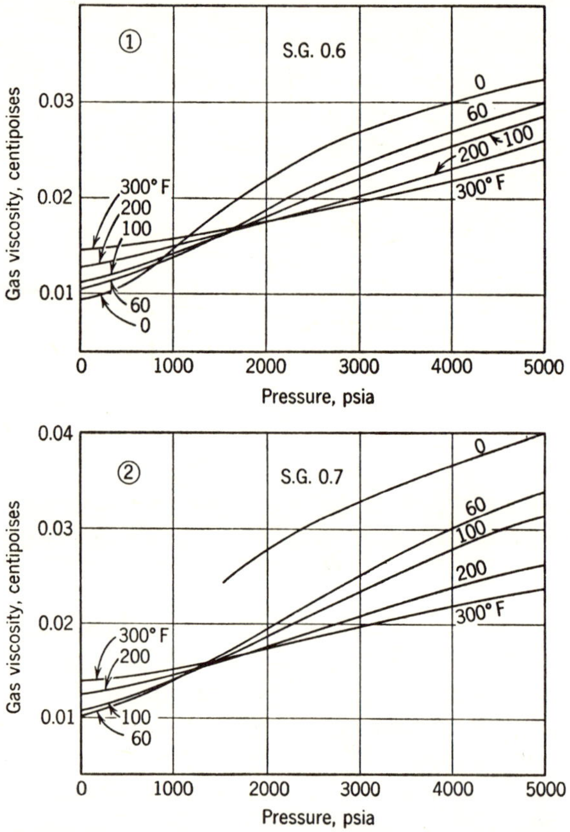

FIG. 79. Viscosity of various specific gravity gases as a function of pressure and temperature. (From Calhoun, John C., Jr., *Fundamentals of Reservoir Engineering,* Copyright 1953, University of Oklahoma Press, pp. 41, 42.)

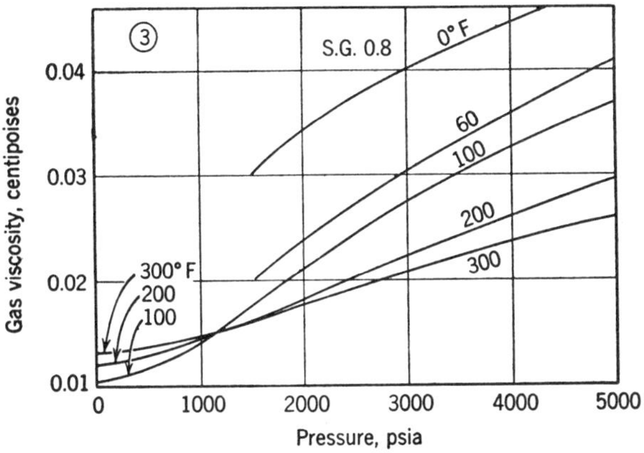

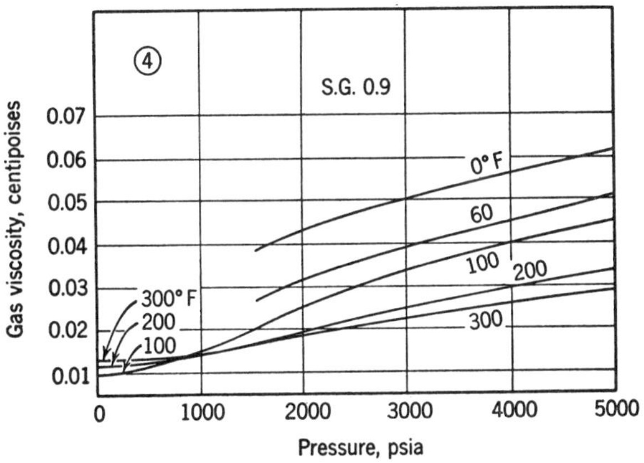

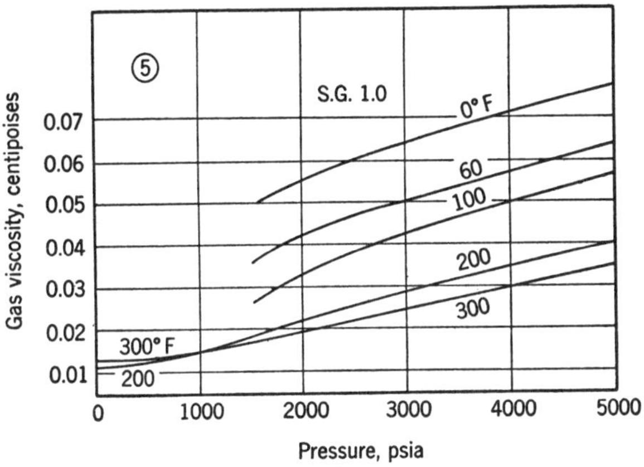

FIG. 79 (*Continued*).

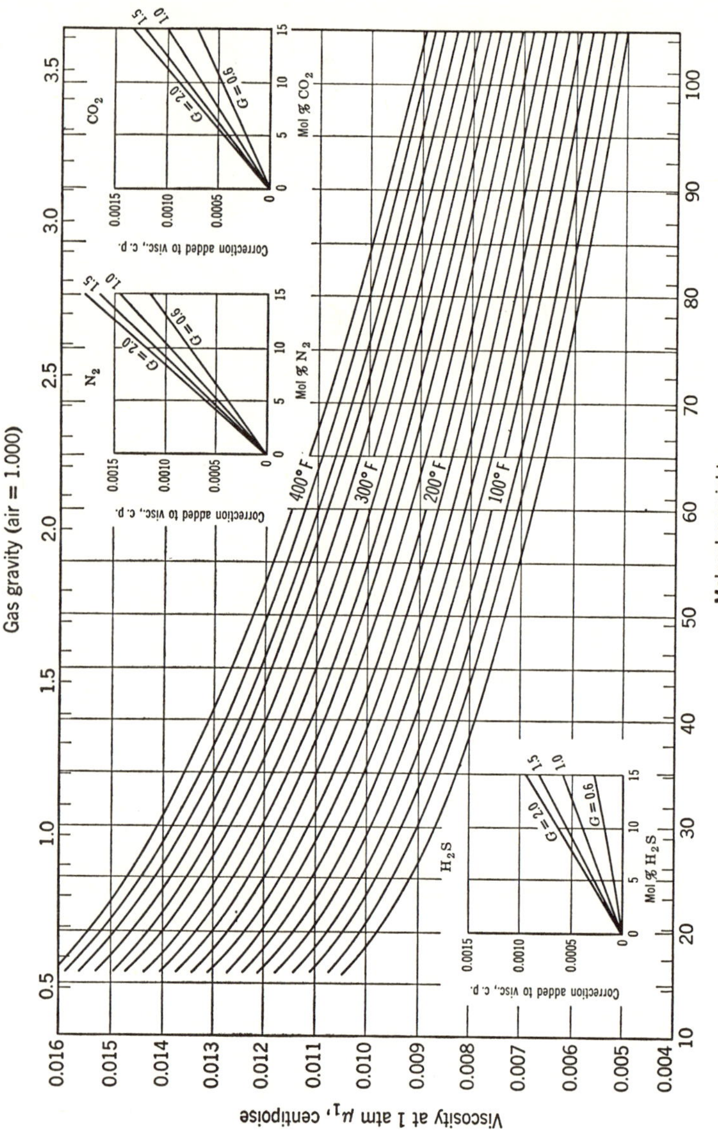

FIG. 80. Viscosity of hydrocarbon gases at one atmosphere pressure. (The symbol G in this chart represents the gas specific gravity.) (Carr, Kobayashi, and Burrows, *Journal of Petroleum Technology*, 1954.)

and μ_1 is the viscosity of the gas at one atmosphere pressure and at reservoir temperature. It has been found that this ratio at a given reduced pressure and temperature is the same for all hydrocarbon gases. Furthermore, the same relationship appears to be true for a hydrocarbon gas mixture provided the pseudo-critical pressure and

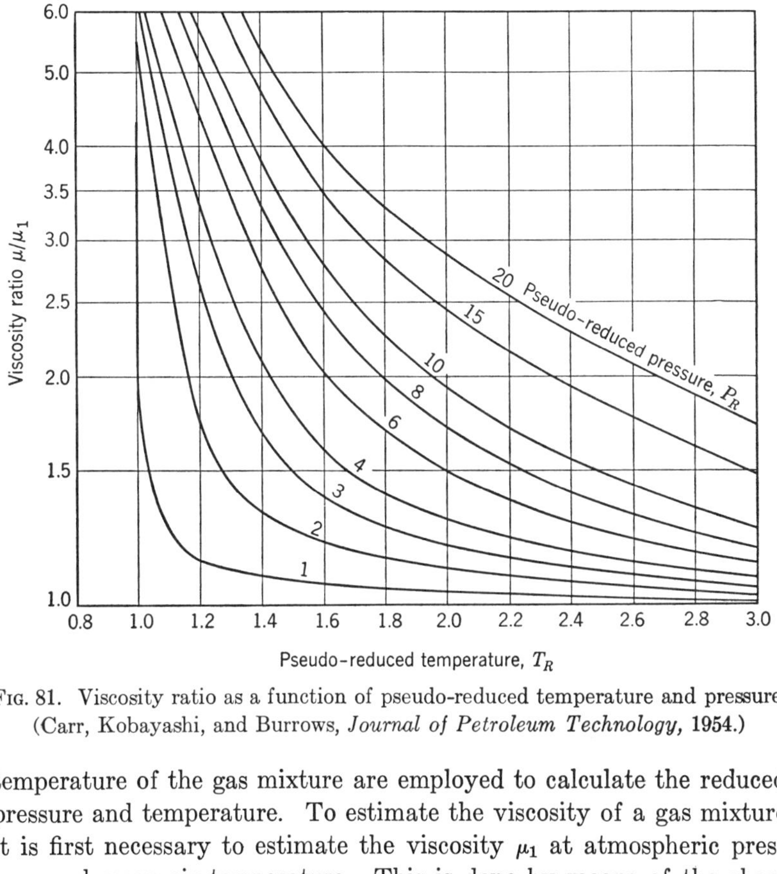

Fig. 81. Viscosity ratio as a function of pseudo-reduced temperature and pressure. (Carr, Kobayashi, and Burrows, *Journal of Petroleum Technology,* **1954.**)

temperature of the gas mixture are employed to calculate the reduced pressure and temperature. To estimate the viscosity of a gas mixture it is first necessary to estimate the viscosity μ_1 at atmospheric pressure and reservoir temperature. This is done by means of the chart shown in Figure 80 which gives the viscosity of gas mixtures at one atmosphere pressure as a function of apparent molecular weight (or gas gravity) and temperature. The pseudo-critical pressure and temperature are calculated from the gas composition in the usual manner (equations 20 and 21, Chapter 2). The viscosity ratio μ/μ_1 at the required pseudo-reduced pressure and temperature is obtained from Figure 81 and μ computed for reservoir conditions.

EXAMPLE. A gas mixture is composed of three pound-moles of methane and one pound-mole of ethane. Estimate the gas viscosity at 2730 psia and 200° F.

The apparent molecular weight of this mixture is given by

$$\text{AMW} = 0.75 \times 16 + 0.25 \times 30 = 19.5$$

The pseudo-critical pressure and temperature are

$$\underline{P_c} = 0.75 \times 673 + 0.25 \times 712 = 683 \text{ psia}$$

$$\underline{T_c} = 0.75 \times 344 + 0.25 \times 549 = 395° \text{ R}$$

The pseudo-reduced pressure and temperature are in this case equal to

$$P_R = \frac{2730}{683} = 4.0 \qquad T_R = \frac{460 + 200}{395} = 1.67$$

The value of μ_1 is found to be 0.0124 c.p. from Figure 80 and μ/μ_1 is found to be 1.50 from Figure 81. Consequently, the required viscosity at 2730 psia and 200° F is

$$\mu = 1.50 \times 0.0124 = 0.0186 \text{ c.p.}$$

Since this method of estimating gas viscosity is based on the Law of Corresponding States a correction must be applied if the mixture contains appreciable quantities of non-hydrocarbon gases. The corrections which must be added to μ because of the presence of N_2, CO_2, or H_2S in the gas mixtures are shown in the insert plots in Figure 80. Having corrected μ_1 for the non-hydrocarbon constituents, computation of μ is carried out in the same manner as before.

This method of estimating the viscosity of gas mixtures requires a knowledge of the gas composition so that the pseudo-critical pressure and temperature may be computed. If composition data are not available it is still possible to employ this method, with a sacrifice of accuracy however, if the gas gravity is known. In this event the pseudo-criticals may be estimated by means of the specific gravity correlation already discussed in Chapter 2 and shown in Figures 11 and 12. Having obtained the pseudo-reduced temperatures and pressure in this way the estimation of the gas viscosity at reservoir temperature and pressure is carried out in the same manner as before.

FLUID CHARACTERISTICS OF OIL RESERVOIR WATER

Since water is almost invariably found in conjunction with petroleum deposits a description of its properties that are pertinent to petroleum engineering problems will be given in the following sections. Its composition, its ability to dissolve hydrocarbon gases, its reservoir formation volume factor, and its viscosity characteristics under reservoir conditions of temperature and pressure will be described. It should be stated at the outset that experimental data on the fluid characteristics of reservoir water are relatively meager. This lack of information can probably be attributed to the fact that the data which are available indicate that the effects of solution gas on the properties of reservoir water are relatively slight and may be neglected for most engineering calculations. However, this fact is in itself sufficient reason to warrant the discussion of water fluid characteristics in some detail.

Composition of Oil Reservoir Water. Oil reservoir water, which is also termed "connate" or "interstitial" water, almost invariably contains dissolved salts. Indeed, comparisons with sea water show that connate waters have *on the average* a higher salt content than sea water. Many analyses of the chemical content of connate water have shown a wide diversity in the nature and amounts of the dissolved ions. The cations most commonly present are Na^+, K^+, Ca^{++}, and Mg^{++} but Ba^{++}, Li^+, and Fe^{++} may also be present. The anions most commonly present are Cl^-, $SO_4^=$, and HCO_3^- but $CO_3^=$, NO_3^-, Br^-, I^-, and $S^=$ are often important constituents. The composition of a typical Pennsylvania oil reservoir water is shown in Table 8.

Table 8. Composition of a Typical Connate Water

Composition Ion	Connate Water from Well #23 Stover Farm, McKean County, Pa. Parts per million	Sea Water Parts per million
Ca^{++}	13,260	420
Mg^{++}	1,940	1,300
Na^+	31,950	10,710
K^+	650	......
$SO_4^=$	730	2,700
Cl^-	77,340	19,410
Br^-	320	......
I^-	10	......
Total	126,200	34,540

The composition of sea water is also listed in this table for comparison. The composition of connate water is usually expressed in parts per million (milligrams of each constituent ion per liter), grams per liter, or grains per gallon.

The Solubility of Natural Gas in Connate Water. The solubility of a natural gas in interstitial water as a function of temperature and pressure has been determined by Dodson and Standing. These investigators used a 0.655 gravity gas and measured the solubility in pure water and two brine samples. The compositions of the gas and the brines are shown in Table 9.

Table 9. Composition of Natural Gas and Brines Used in the Experimental Gas Solubility Measurements

Composition of Natural Gas		Composition of Brines		
			Parts per million	
Component	Mole Fraction	Component	Brine *A*	Brine *B*
CH_4	0.8851	Na^+	3,160	12,100
C_2H_6	0.0602	Ca^{++}	58.1	520
C_3H_8	0.0318	Mg^{++}	39.8	380
i-C_4H_{10}	0.0046	$SO_4^{=}$	0.0	5.3
n-C_4H_{10}	0.0085	Cl^-	4,680	20,000
i-C_5H_{10} and heavier	0.0098	HCO_3^-	696	980
		I^-	0.0	130
	1.0000		8,633.9	34,115.3

The experimental results are shown in Figure 82. The upper plot represents the solubility of the gas in pure water while the lower plot gives the correction factors necessary to account for the decrease in gas solubility with increasing salinity of the water.

These results permit certain generalizations to be made concerning the solubility of natural gas in water and brine.

1. The solubility of natural gas in connate water is *small* compared to its solubility in crude oil at comparable temperatures and pressures. An examination of Figure 82 shows that the solubility at 2000 psia is of the order of 12 S.C.F. per barrel. At this pressure the solubility of natural gas in crude oil is about 900 S.C.F. per barrel for a 50° API crude. (See Figure 57.)

2. At constant temperature the solubility of natural gas in connate water increases with pressure. However, the solubility is not a linear function of pressure as required by Henry's Law.

3. At constant pressure the solubility initially decreases with temperature. However, at high pressures the solubility reaches a minimum so that a further increase in temperature brings about an increase in solubility. As already pointed out this effect is not observed in natural-gas–crude-oil systems in the temperature and pressure ranges investigated. However, these minima in solubility are not uncommon for gases which are only slightly soluble in a liquid.

4. The solubility of natural gas in brine decreases with increasing salinity of the brine.

5. When the results shown in Figure 82 are compared with solubility data for pure ethane in water one reaches the conclusion that the

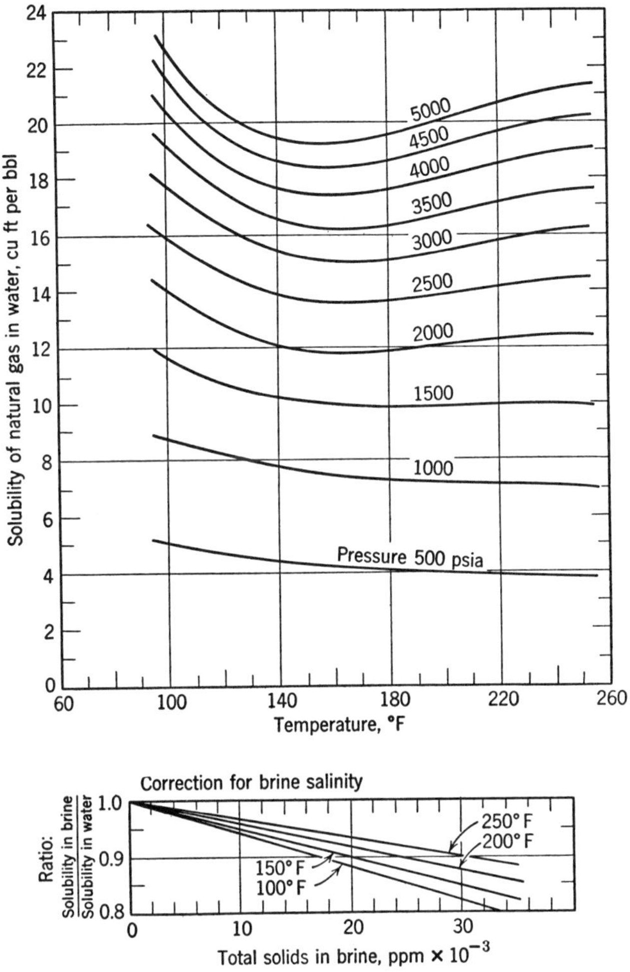

FIG. 82. Solubility of natural gas in water. (Dodson and Standing, *API Drilling and Production Practice,* 1944, p. 173.)

solubility of natural gas in connate water decreases with increasing gas gravity.

Figure 82 may be used to estimate the solubility of a natural gas in a brine. The accuracy of the estimate will naturally depend on how closely the gas and brine in question resemble those shown in Table 9.

EXAMPLE. Estimate the solubility of a natural gas at a pressure of 2000 psia and a temperature of 150° F in a brine containing 20,000 parts per million of dissolved solids.

From the upper plot of Figure 82 the solubility of the natural gas in water at 2000 psia and 150° F is found to be 11.9 S.C.F. per barrel. From the lower plot the correction factor for brine salinity at 150° F and 20,000 parts per million total solids is 0.9. Therefore, the required solubility is $11.9 \times 0.9 = 10.7$ S.C.F. per barrel.

The Water-Formation Volume Factor. A typical plot of β_w as a function of pressure is shown in Figure 83. As the pressure is de-

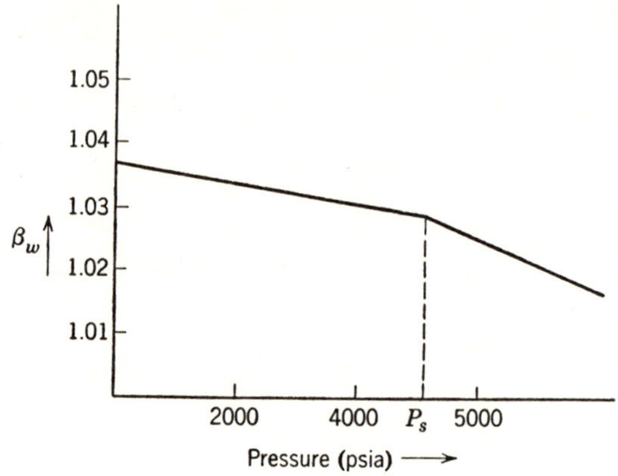

FIG. 83. Typical plot of the water-formation volume factor as a function of pressure at reservoir temperature.

creased from P_0 to P_s the value of β_w increases due to the expansion of the liquid. At pressures below P_s gas is evolved, but because of the low solubility of natural gas in brine, the shrinkage of the liquid phase is relatively small. This shrinkage is usually insufficient to counterbalance the expansion of the liquid on release of pressure so that β_w continues to *increase* below the saturation pressure as shown. However, the rate of increase with decreasing pressure is lower at pressures below P_s than at pressures above P_s. It is also noteworthy that in the pressure ranges most commonly encountered in petroleum reservoirs the value of β_w does not differ greatly from one. This also is a consequence of the low solubility of natural gas in brine.

Experimental values of the water-formation volume factor below the saturation pressure are shown in Figure 84. This chart was prepared from data obtained by Dodson and Standing using the natural gas whose composition is given in Table 9. The upper curve of each pair shown in Figure 84 represents β_w as a function of pressure at constant temperature for *pure water containing dissolved natural gas.* The lower curve of each pair represents the formation volume factor

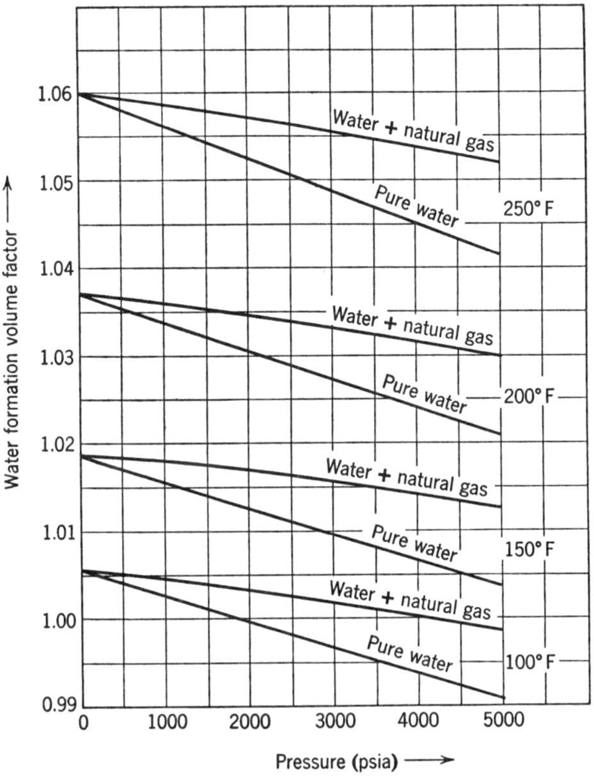

FIG. 84. Water-formation volume factor for pure water and mixtures of water and natural gas. Factors are plotted as a function of pressure at 100°, 150°, 200°, and 250° F.

of *pure water* as a function of pressure at constant temperature. The increase in β_w with decreasing pressure is clearly shown in this diagram, as is the fact that β_w is only slightly different from one in the pressure range investigated.

Figure 84 may be used to estimate a value of β_w below the saturation pressure. If the accuracy of other data on the reservoir justifies an additional refinement, the value of β_w for brines of various total salt concentration can be estimated provided a correction is made for the decreased gas solubility in brine as is shown in the example below.

EXAMPLE. Estimate β_w for a brine containing 20,000 parts per million of total solids at 2000 psia and 150° F.

In the previous example it was shown that the solubility of gas in pure water and in 20,000 ppm brine was 11.9 and 10.7 S.C.F. per barrel respectively at 2000 psia and 150° F. From Figure 84 it is evident that the increase in β_w caused by the solution of 11.9 S.C.F. of gas is 1.017 − 1.013 or 0.004. Since only 10.7 S.C.F. of gas dissolve in the brine the increase in β_w in this case would only be

$$\frac{10.7}{11.9} \times 0.004 = 0.0036$$

Consequently, the required value of β_w is

$$1.013 + 0.0036 = 1.0166$$

The values of β_w at pressures above the saturation pressure are determined by the coefficient of compressibility of the gas-saturated connate water. Value of these coefficients are given in Figure 85. The upper plot gives the coefficient of compression of pure water or a gas-free brine as a function of temperature and pressure. The lower plot gives the multiplicative correction factor which must be applied to account for the increase in compressibility due to solution gas.

EXAMPLE. Estimate the coefficient of compression of a 20,000 ppm total solids brine saturated with gas at 2000 psia and 150° F.

From the upper plot in Figure 85 the value of the coefficient of compression of a gas-free brine is found to be 3.0×10^{-6} bbl per bbl per psi. Since the gas solubility is 10.7 S.C.F. per bbl (see preceding examples) the correction factor is found to be 1.1 in the lower plot of Figure 85. Consequently, the required coefficient is

$$3.0 \times 10^{-6} \times 1.1 = 3.3 \times 10^{-6} \text{ bbl per bbl per psi}$$

It should be apparent that the coefficient of compression of a given gas-saturated water may be estimated at any temperature and pressure with the aid of the data in Figure 85. Consequently, the *average* coefficient may be estimated for any pressure range above the saturation pressure. Values of β_w above the saturation pressure may then be computed using the equation

$$\beta_w = \beta_{w_s}[1 - C_{avg}(P - P_s)]$$

where β_w is the value of the water-formation volume factor above the saturation pressure at pressure P, β_{w_s} is the value of the water-formation volume factor at the saturation pressure P_s, and C_{avg} is the average coefficient of compression over the pressure range P to P_s.

The data presented in this section show the effects of solution gas on the reservoir water volume. In general, due to the low solubility of natural gas in connate water, these effects are small. Consequently, in the next chapter, when reservoir equations involving reservoir fluid characteristics are discussed on an elementary basis, the volume changes of the water phase and the solubility of gas in connate water will be neglected.

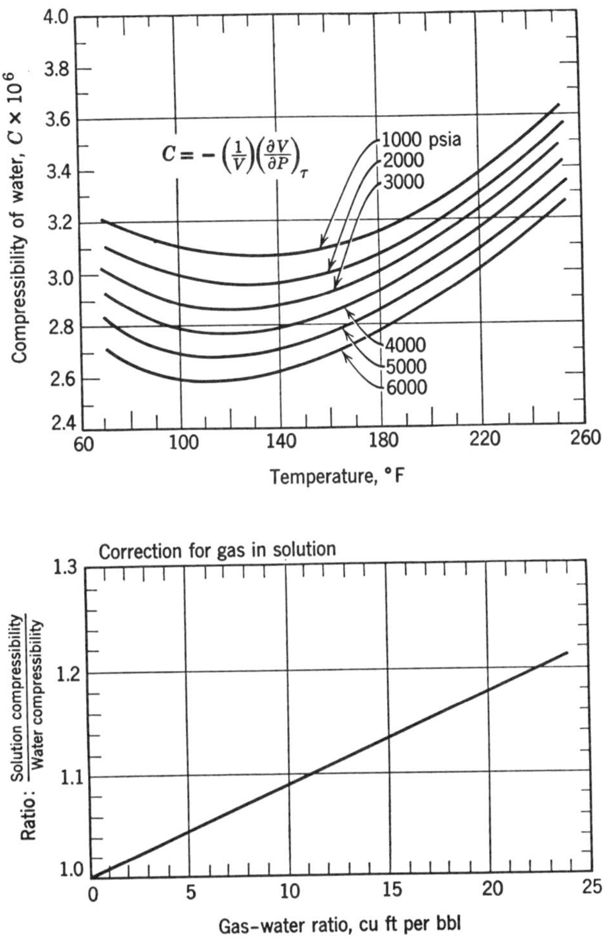

FIG. 85. Effect of dissolved gas on the compressibility of water. (Dodson and Standing, *API Drilling and Production Practice,* 1944, p. 173.)

The Viscosity of Connate Water. The effect of high pressures on the viscosity of pure water was investigated by Bridgeman in 1926. Some of his results are reproduced in Table 10. From an examination of the data presented in Table 10 it is immediately apparent that the change in water viscosity with pressure is small for the pressure ranges usually encountered in petroleum engineering practice.

Table 10. Viscosity of Pure Water as a Function of Pressure at Various Temperatures

Pressure psia	32° F	50.5° F	86° F	167° F
14.22	1.79 c.p.	1.40 c.p.	0.876 c.p.	0.398 c.p.
7110	1.68 c.p.	1.35 c.p.	0.897 c.p.	0.413 c.p.
14,220	1.65 c.p.	1.33 c.p.	0.922 c.p.	0.429 c.p.

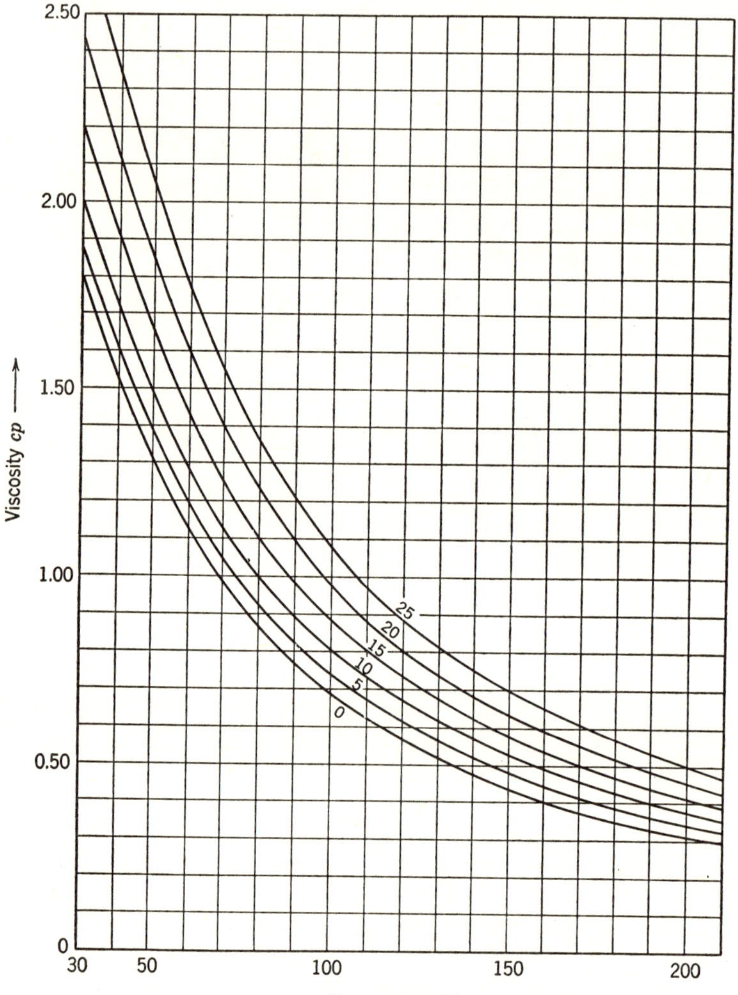

FIG. 86. Viscosity of sodium chloride solutions as a function of temperature for solutions containing 0, 5, 10, 15, 20, and 25 grams of sodium chloride per 100 grams of water. (At atmospheric pressure.)

The addition of salts to water causes an increase in viscosity as is shown in Figure 86. The data presented are for sodium chloride solutions containing 0, 5, 10, 15, 20 and 25 grams of the salt per 100 grams of water. This figure also shows the dependence of viscosity on temperatures for these solutions.

It has been pointed out that small amounts of natural gas dissolve in connate water at reservoir temperature and pressure. However, no data on the effect of this solution gas on water viscosity have been published. Undoubtedly, the solution gas causes a decrease in the water viscosity but the magnitude of this decrease is unknown.

In view of the lack of experimental data it is impossible to make an accurate estimate of connate water viscosity under reservoir conditions. Consequently, if an experimental value for the brine viscosity is not known, it is common practice to assume a value equal to that of pure water at atmospheric pressure and at reservoir temperature (see curve for 0% sodium chloride in Figure 86). This tacitly assumes that the viscosity of brine is independent of pressure and that the increase in viscosity caused by dissolved salts is more or less counterbalanced by the probable decrease caused by solution gas.

EXPERIMENTAL DETERMINATION OF RESERVOIR FLUID CHARACTERISTICS

In order to give the reader a more complete understanding of the reservoir fluid characteristics just described, an outline of the experimental techniques and procedures employed for their measurement will be presented. It should be stated at the outset that many variations in these techniques exist and each laboratory follows its own detailed procedure. Moreover, the precise experimental determination of the various reservoir fluid characteristics at high pressures and relatively high temperatures is a rather involved process. Consequently, the material to be presented should not be considered as a complete discussion of the subject. Rather, it should be considered as an outline of the fundamental principles involved.

Determination of r and β. Samples of reservoir oil for analysis are obtained either by recombining surface samples of oil and gas in the proper proportions or by obtaining a sample under reservoir conditions by means of a bottom hole sampler.

Samples obtained in this manner are transferred at pressures above the bubble point to a heavy-walled, stainless-steel vessel capable of withstanding high pressures. The volume of this vessel, which is known as a pressure-volume-temperature (*PVT*) cell, may be varied by injecting or withdrawing mercury through an inlet tube situated at the base of the cell. The accessory apparatus consists of a mercury pump for applying pressure and for injecting or withdrawing *known* volumes of mercury in the cell, a wet-test meter or other gas-measuring device for determining the volume of gas in solution, and a

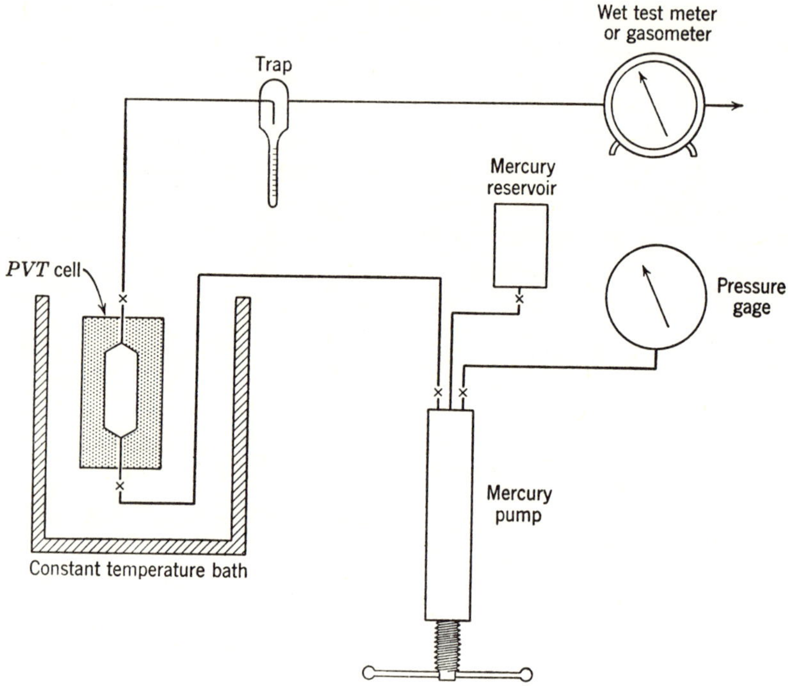

FIG. 87. Diagrammatic representation of apparatus for the determination of reservoir fluid characteristics.

constant temperature bath for maintaining the *PVT* cell and its contents at reservoir temperature. A diagrammatic representation, showing the arrangement of the apparatus, is shown in Figure 87.

Before measurements of r and β as a function of pressure are made it is necessary to determine the saturation pressure. This pressure is readily determined by observing the change in pressure within the cell when mercury is injected into the *PVT* cell from the mercury pump. At pressures above the bubble point the injection of a given amount of mercury causes a large increase in pressure when compared to the increase in pressure caused by the injection of the same amount of mercury at pressures below the bubble point. A typical plot of volume of mercury injected versus pressure is shown in Figure 88. The sharp break which occurs in the curve at the saturation pressure is clearly evident.

As previously pointed out, values of r and β depend on the manner in which the gas is liberated. In the procedure to be described the pressure is reduced in 200 psi steps from the bubble point to atmospheric pressure and the gas bled off after each step. The values of r

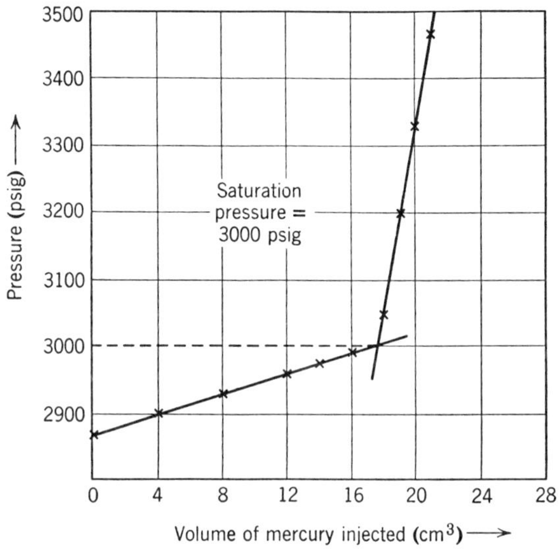

Fig. 88. Determination of saturation pressure.

and β determined in this manner will *approach* those for true differential liberation. Consequently, in laboratory parlance it is customary to designate this experimental technique as a differential liberation. In the laboratory, flash liberations are generally conducted over much larger pressure drops.

For the determination of the differential values of r and β the *PVT* cell containing mercury and a known volume of reservoir fluid is immersed in the constant temperature bath at reservoir temperature and the pressure is reduced 200 psi below the saturation pressure by withdrawing mercury from the cell through the mercury pump. The cell and its contents are thoroughly agitated until equilibrium is established and the volume of the gas-oil system is recorded. The gas is bled off through the metering device and at the same time the piston of the mercury pump is slowly advanced to keep the pressure in the cell constant. When the gas has been bled off, the volume of the residual oil in the cell is measured and recorded. The volume of the evolved gas is measured and corrected to standard conditions. The pressure is decreased in steps of 200 psi and the process repeated until atmospheric pressure is reached. The cell is removed from the constant temperature bath and the residual oil volume is determined and corrected to stock tank conditions. At each pressure the value of β is given by the ratio of the oil volume at reservoir conditions to the oil volume at stock tank conditions. Values of r are obtained by computing the number of standard cubic feet of gas dissolved per barrel of stock tank oil at each pressure. Values of r and β calculated in this manner are shown in Table 11. The experimental data pre-

Table 11. Computation of Reservoir Fluid Characteristics

(1) Pressure (psig)	(2) Volume of Oil and Gas at Reservoir Conditions (cc)	(3) Volume of Oil at Reservoir Conditions (cc)	(4) Volume of Gas at Reservoir Conditions, column 2 − column 3 (cc)	(5) Total Gas Evolved, corrected to standard conditions (cc)	(6) β $\frac{\text{Column 3}}{56.02}$	(7) Gas in Solution, 6,500 − column 3 (cc)	(8) r S.C.F./S.T.B. $\frac{\text{column 7} \times 5.62}{56.02}$	(9) Z
3000	.....	75.00	....	0	1.339	6500	652.1	.....
2800	75.95	74.21	1.74	370	1.325	6130	615.0	0.732
2600	75.26	73.39	1.87	750	1.310	5750	576.8	0.711
2400	74.55	72.60	1.95	1,120	1.296	5380	539.7	0.703
2200	73.98	71.80	2.18	1,500	1.282	5000	501.6	0.702
2000	73.44	71.01	2.43	1,870	1.268	4630	464.5	0.731
1800	73.09	70.21	2.88	2,250	1.253	4250	426.4	0.760
1600	72.63	69.40	3.23	2,620	1.239	3880	389.2	0.779
1400	72.45	68.61	3.84	2,990	1.225	3510	352.1	0.812
1200	72.54	67.79	4.75	3,370	1.210	3130	314.0	0.839
1000	72.75	67.00	5.75	3,740	1.196	2760	276.9	0.871
800	73.73	66.11	7.62	4,125	1.180	2375	238.3	0.891
600	77.57	64.72	12.85	4,600	1.155	1900	190.6	0.919
400	83.71	63.01	20.70	5,100	1.125	1400	140.4	0.949
200	106.31	60.81	45.50	5,650	1.085	850	85.3	0.982
0	.....	58.20		6,500	1.039	0	0	1.000
0 (60° F)	.....	56.02						

Saturation pressure = 3000 psig.
Reservoir temperature = 180° F.

sented in this table are for an oil with a saturation pressure of 3000 psig taken from a reservoir whose temperature is 180° F. These data have been simplified for purposes of illustration. For example, in columns 2 and 3 the fluid volumes measured under reservoir conditions have already been corrected for compression and thermal expansion of mercury and for changes in cell volume due to changes in pressure and temperature. Similarly, the gas volumes recorded in column 5 have been corrected to standard conditions. The values of β and r have been calculated as indicated and tabulated in columns 6 and 8,

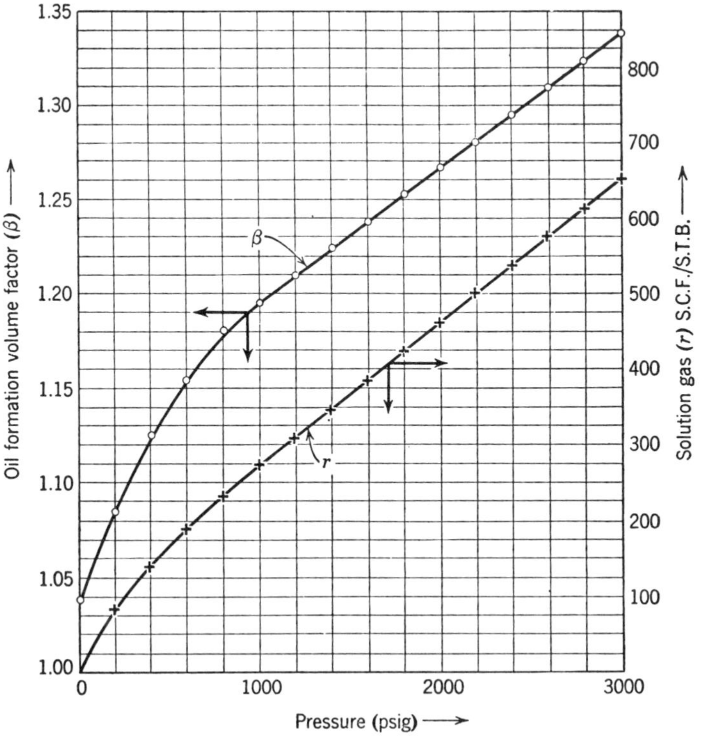

FIG. 89. r and β as a function of pressure. (Data from Table 11.)

respectively. Plots of these values as a function of pressure are shown in Figure 89.

Determination of the Z Factor. The data obtained in the determination of r and β may also be used to evaluate the compressibility factor Z (and consequently v) as a function of pressure at reservoir temperature. Since the volume of the gas phase is known both under reservoir conditions and standard conditions the value of Z may be computed using the equation

$$\frac{P_1V_1}{Z_1T_1} = \frac{P_0V_0}{Z_0T_0}$$

where the subscript 1 represents standard conditions and the subscript 0 represents reservoir conditions. Since $Z_1 = 1$ this equation may be solved for Z_0 giving

$$Z_0 = \frac{P_0V_0T_1}{P_1V_1T_0} \tag{10}$$

In this equation V_0 is given by the difference between the pump reading when the system consists of both oil and gas and the pump reading when all the gas has been bled off and only oil remains. In Table 11 V_0 in cm^3 is given in column 4 and is equal to the difference between column 2 and column 3. V_1 in this equation is given in cm^3 by the difference between succeeding values in column 5. Since the pressures and temperatures are known Z_0 in equation 10 may be calculated.

EXAMPLE. Using the data in Table 11 calculate Z_0 at 2600 psig and 180° F

$$P_0 = 2614.7 \text{ psia}, P_1 = 14.7 \text{ psia}, T_0 = 640° \text{ R}, T_1 = 520° \text{ R}$$

$$V_0 = 75.26 - 73.39 = 1.87 \text{ cm}^3 \quad \text{and} \quad V_1 = 750 - 370 = 380 \text{ cm}^3$$

Consequently, substitution in equation 10 gives the following value for Z_0 at 2600 psig and 180° F

$$Z_0 = \frac{2614.7 \times 1.87 \times 520}{14.7 \times 380 \times 640} = 0.711$$

Values of Z_0 calculated in this manner are given in column 9 in Table 11 and a plot of Z_0 versus pressure is shown in Figure 90.

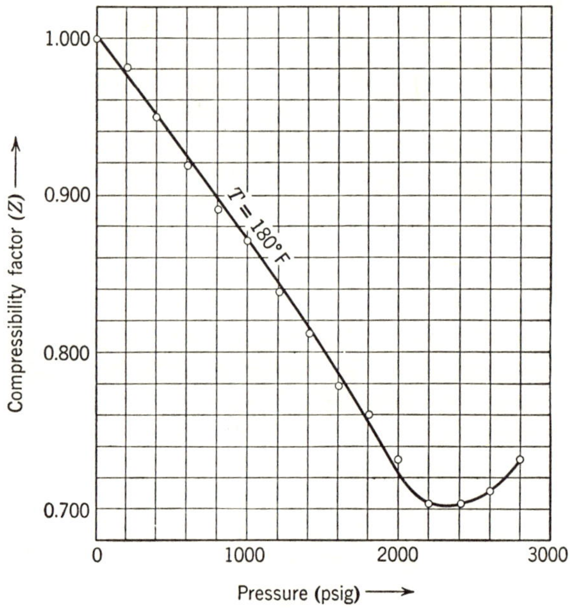

FIG. 90. Compressibility factor as a function of pressure. (Data from Table 11.)

Determination of Fluid Viscosities. Although many experimental methods are available for determining fluid viscosities, only a few are readily adaptable to measurements at high pressures and relatively high temperatures. Under these conditions, a rolling ball viscosimeter may be employed to measure the viscosities of both liquids and gases. Since a Rankine capillary viscosimeter is also adaptable to high-pressure gas viscosity measurements this method is sometimes employed for this purpose.

A rolling ball viscosimeter consists essentially of a cylindrical tube which is inclined at a definite angle. The tube is filled with the fluid whose viscosity is to be measured and a metal ball is allowed to roll down through the tube. The bottom of the tube is closed so that as the ball travels downward the fluid passes upward through the space between the rolling ball and the walls of the tube. The time for the descent of the ball is accurately measured. It may be shown that the velocity of the ball is given by

$$V = \text{constant} \frac{(D - d)}{\mu} \tag{11}$$

where the constant is determined by the dimensions of the instrument, D is the density of the ball, d is the density of fluid, and μ is the viscosity of the fluid. When used as a relative method the instrument is calibrated using a fluid of known viscosity. Under these conditions equation 11 becomes

$$\frac{\mu_2}{\mu_1} = \frac{(D - d_2)t_2}{(D - d_1)t_4} \tag{12}$$

where t denotes the time of fall through a fixed distance and the subscripts 1 and 2 refer to the standard fluid and the unknown fluid respectively.

To adapt this instrument for the measurement of viscosities of crude oils with gas in solution at high pressures and temperatures, it is arranged as shown in Figure 91. The tube in this case is constructed of steel and is about 8 in. long and has an internal diameter of about ¼ in. It fits in a slightly larger hole bored in a heavy steel cylinder which is capable of withstanding high pressures. This heavy cylinder is mounted on a trunnion so that it may be rotated through an angle of about 330°. The trunnion is fitted with a stop which when in place gives the instrument a definite angle of inclination of 75° from the horizontal. The oil and gas are introduced through the inlet tube into the enlarged space at the top of the instrument. This enlarged space allows the fluid to be properly agitated by rocking the entire apparatus in a constant temperature bath until equilibrium between the gas and oil is established. During this process the ball is kept out of the tube by a retractable plunger which passes through the cylinder head. When equilibrium has been established the ball is allowed to fall into place by retracting the plunger. Then the plunger is screwed down on the top of the tube so that it seals the upper end

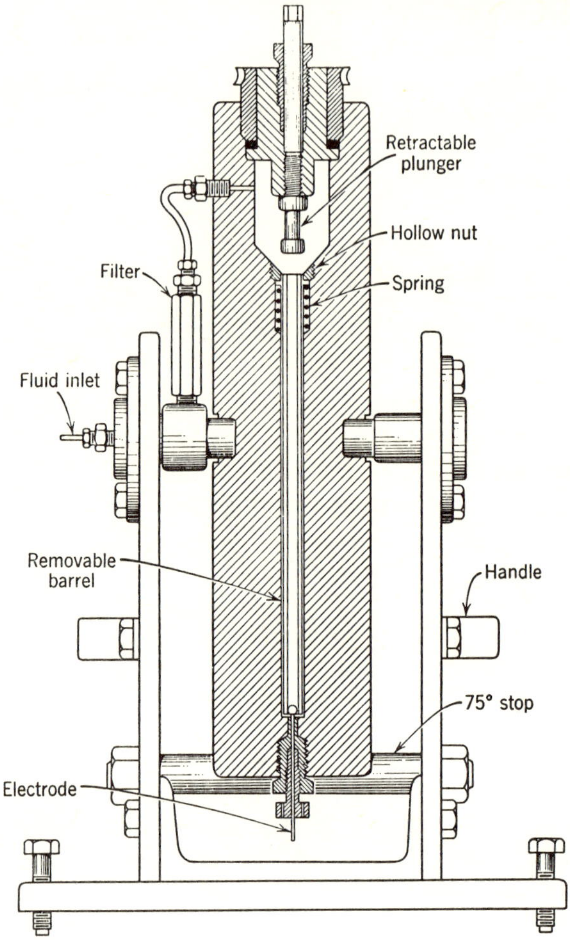

FIG. 91. Diagram of rolling ball pressure viscosimeter. (Hocott and Buckley, *AIME Trans.*, *142*, 1941.)

and simultaneously presses the bottom onto a gasket so that the lower end of the tube is also sealed.

To make a viscosity determination the instrument is rotated about 180° so that the ball falls to the top of the instrument. The instrument is then rotated rapidly back against the stop. When the metal ball reaches the bottom of the tube it makes electrical contact with an electrode and actuates a signal so that the time of fall can be determined. By repeating this procedure with a fluid of known viscosity the viscosity of the unknown fluid may be calculated using equation 12.

When the rolling ball viscosimeter is employed to measure gas viscosities it is necessary to use a ball which just fits the tube. Under these conditions the rate of fall is slow enough so that it can be accurately measured. Notice that for gases at *low pressures* equation 12 reduces to

$$\frac{\mu_2}{\mu_1} = \frac{t_2}{t_1}$$

since d_1 and d_2 are negligibly small compared to D in this case.

A second method for the measurement of gas viscosity at high pressures and temperatures involves the use of a Rankine capillary vis-

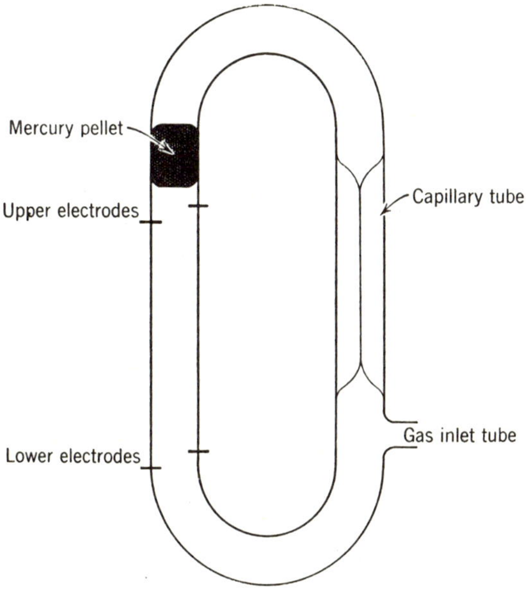

FIG. 92. Diagrammatic representation of a Rankine capillary viscosimeter.

cosimeter. This viscosimeter consists of two parallel glass tubes joined at both ends as shown in Figure 92. One tube is a heavy-walled capillary and the other is a regular thin-walled tube about ¼ in. internal diameter. A mercury pellet in the large tube falls under the action of gravity and displaces the gas through the capillary.

The flow of fluid through a capillary tube is governed by Poiseuille's

Law. For viscous flow this law in mathematical form may be written

$$\mu = \frac{\pi r^4 Pt}{8LV} \tag{13}$$

where μ is the viscosity in poises, r is the radius of the capillary in cm, P is the pressure differential causing flow in dynes/cm^2, t is the time in seconds for a volume V in cm^3 to flow through the tube and L is the length of the tube in cm. In the case of a Rankine viscosimeter, P is the constant pressure exerted by the mercury pellet as it falls at a uniform rate in the larger tube.

To adapt the Rankine viscosimeter for high-pressure measurements the entire apparatus is enclosed in a steel bomb. The bomb is filled with nitrogen gas at a pressure near that of the gas contained in the viscosimeter. When the instrument is thus enclosed it is necessary to measure the fall time of the mercury pellet electrically. This is accomplished by sealing in two electrodes near the top of the fall tube and two more near the bottom. When the falling mercury pellet contacts these electrodes it actuates a signal. The entire bomb is enclosed in a constant temperature bath and is arranged so that it may be rotated about an axis perpendicular to the fall tube. In this way the mercury pellet can be brought to the top of the fall tube. To make a viscosity determination the mercury pellet is brought to a point above the upper pair of electrodes. The instrument is rapidly rotated to a vertical position and the fall time determined. This method may be used either as a relative method or as an absolute method. In the former case the instrument is calibrated with gases of known viscosity. In the latter case it is necessary to know the dimensions of the apparatus so that the gas viscosity may be computed using equation 13.

REFERENCES

Bicher, L. B., and D. L. Katz, Viscosity of Natural Gases, *Trans. AIME, 155,* 244 (1944).

Beal, C., Viscosity at Oil Field Temperatures and Pressures, *Trans. AIME, 165,* 94 (1946).

Calhoun, J. C., *Fundamentals of Reservoir Engineering,* University of Oklahoma Press, Norman, Okla. (1953).

Carr, N. L., R. Kobayashi, and D. B. Burrows, Viscosity of Hydrocarbon Gases under Pressure, *J. Petroleum Technol.,* Oct. 1954.

Dodson, C. R., and M. B. Standing, Pressure-Volume-Temperature and Solubility Relations for Natural-Gas–Water Mixtures, *API Drilling and Production Practice,* p. 173, 1944.

Hocott, C. K., and S. E. Buckley, Measurements of the Viscosities of Oils under Reservoir Conditions, *Trans. AIME, 142,* p. 131 (1941).

Katz, D. L., Prediction of Shrinkage of Crude Oils, *API Drilling and Production Practice,* p. 137, 1942.
Muskat, M., *Physical Principles of Oil Production,* McGraw-Hill Book Co., New York (1949).
Pirson, S. J., *Elements of Oil Reservoir Engineering,* McGraw-Hill Book Co., New York (1950).
Standing, M. B., A Pressure-Volume-Temperature Correlation for Mixtures of California Oils and Gases, *API Drilling and Production Practice,* p. 275, 1947.
Standing, M. B., *Volumetric and Phase Behavior of Oil Field Hydrocarbon Systems,* Reinhold Publishing Corp., New York (1952).

PROBLEMS

1. A gas has a specific gravity of 0.740. Calculate Z and then v at 210° F and 2300 psia. Compare this value of v with that obtained directly from charts in Figure 51

2. An oil has a formation volume factor of 1.34. What is the shrinkage factor, the shrinkage based on reservoir oil volume, and the shrinkage based on S.T.O. volume?

3. A 45° API oil dissolves 825 S.C.F. per barrel of 0.74 gravity gas at 2000 psia and 150° F. The composition of the gas is as follows:

Component	Volume %
CH_4	81.0
C_2H_6	7.5
C_3H_8	5.5
C_4H_{10}	4.0
C_5H_{12}	1.5
C_6H_{14}	0.5

Estimate β by the four methods described in this chapter and compare results.

4. What is the specific gravity of a crude oil at the saturation pressure if $r = 800$ S.C.F., the gas gravity is 0.9, the oil gravity is 43° API and β_s is 1.5? Estimate the average coefficient of compression above the saturation pressure. *Answer:* S.G. = 0.646.

5. An all-liquid reservoir sample whose original volume was 310.0 cc under reservoir conditions was cooled to 60° F and the pressure released to one atmosphere. The liquid volume was reduced to 204.0 cc and 0.77 S.C.F. of gas were evolved. Calculate r and β. *Answer:* $\beta = 1.52$, $r = 600$ S.C.F./S.T.B.

6. 250.0 cc of a reservoir sample of oil is placed in a PVT cell at 2300 psia and 200° F. On reducing the pressure below the saturation pressure the volume was found to be 264.0 cc. After bleeding off the gas the liquid volume was 241.0 cc. On reducing the pressure and temperature to standard conditions the volume of S.T.O. was found to be 176.0 cc. Calculate β_0, β_1, u_0 and u_1.

7. A PVT cell contains 320 cc of oil and solution gas in a single phase at the saturation pressure of 2500 psia and 200° F. When the pressure was reduced to 2000 psia the volume increased to 335.2 cc. The gas was bled off and found to occupy a volume of 0.145 S.C.F. The volume of oil was 303.0 cc. The pressure was reduced to 14.7 psia and the temperature to 60° F while 0.580 S.C.F. of gas was evolved leaving 230 cc of oil. Calculate r, β, v, and u at 2000 psia. *Answer:* $\beta = 1.32$, $u = 1.46$, $v = 0.0014$ Bbl/S.C.F.

8. Using the data from problem 7 calculate Z at 2000 psia and 200° F. *Answer:* $Z = 0.840$.

9. At 1800 psia and 150° F calculate u from the following data. At the original reservoir pressure of 2000 psia r_0 is 680 S.C.F./S.T.B. At 1800 psia r is 500 S.C.F./S.T.B. At 1800 psia and 150° F the gas compressibility factor is 0.66 and the shrinkage based on original reservoir oil volume is 0.212.

10. A 20° API oil has a saturation pressure of 2000 psia. If the reservoir temperature is 100° F estimate the viscosity from 0 to 2000 psia at intervals of 500 psi and construct a viscosity versus pressure curve.

11. For the natural gas whose composition is given in problem 3 above estimate its viscosity at 2000 psia and 200° F.

12. An interstitial water has a salinity of 30,000 ppm. Estimate the solution gas and the water-formation volume factor at 4000 psia and 200° F.

13. A system at 60° F contains one mole of propane and one mole of n-octane. Assuming ideal-solution behavior calculate: (a) r at 60° F for a flash liberation from the bubble point to 14.7 psia; (b) β at the bubble point.

CHAPTER 7

ELEMENTARY APPLICATIONS OF RESERVOIR FLUID CHARACTERISTICS

To aid the reader to understand the full significance of the reservoir fluid characteristics presented in the previous chapter, they will now be used to develop several important reservoir equations. These equations are frequently used in reservoir analysis calculations to predict the behavior of a reservoir at any time in its future life.

Material Balance Equation for a Constant Volume Reservoir and with No Initial Gas Cap. Reservoir material balances are based on the principle of conservation of mass which asserts that the total mass of a system remains constant during a chemical or physical change. Before applying this principle to a *constant volume reservoir* with *no gas cap* the following terms need to be defined.

N = barrels of stock tank oil originally in reservoir

ΔN = barrels of stock tank oil produced

V_g = standard cubic feet of gas produced

$R_c = V_g/\Delta N$ = cumulative gas-oil ratio

The terms r_0, r, β_0, β, and v have their usual meaning in the following development.

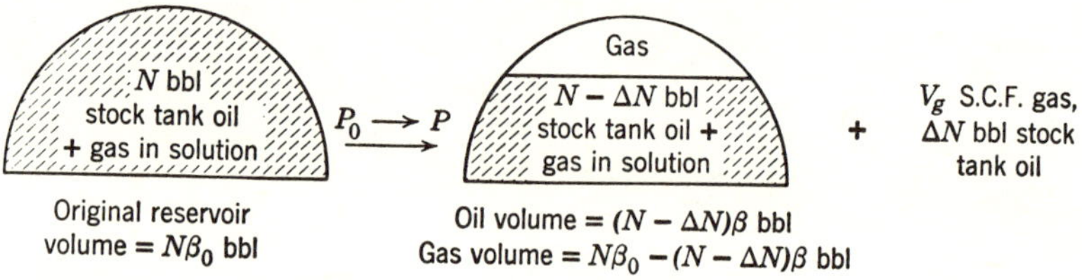

FIG. 93. Volume relationships for a production interval from a constant volume reservoir with no initial gas cap.

Consider the process shown in Figure 93. A material balance on the *gas* after an interval of production may be written

$$\begin{array}{l} \text{S.C.F. of gas} \\ \text{originally in} \\ \text{solution} \end{array} = \begin{array}{l} \text{S.C.F. of free} \\ \text{gas which is} \\ \text{formed } \textit{after} \text{ a} \\ \text{production} \\ \text{interval} \end{array} + \begin{array}{l} \text{S.C.F. of gas} \\ \text{remaining in} \\ \text{solution} \end{array} + \begin{array}{l} \text{S.C.F.} \\ \text{of gas} \\ \text{produced} \end{array} \qquad (1)$$

A little consideration will show that each term in equation 1 can be written in terms of the reservoir fluid characteristics as indicated below.

$$\text{S.C.F. of gas originally in solution} = Nr_0$$

$$\text{S.C.F. of free gas in gas cap} = \frac{N\beta_0 - (N - \Delta N)\beta}{v}$$

$$\text{S.C.F. of gas remaining in solution} = (N - \Delta N)r$$

$$\text{S.C.F. of gas produced} = V_g = R_c\Delta N$$

Consequently, equation 1 becomes

$$Nr_0 = \frac{N\beta_0 - (N - \Delta N)\beta}{v} + (N - \Delta N)r + R_c\Delta N$$

which on rearrangement gives

$$N[\beta + (r_0 - r)v - \beta_0] = \Delta N[\beta - rv + R_cv] \qquad (2)$$

Equation 2 can be simplified by adding and subtracting ΔNr_0v from the right-hand side to give

$$N[\beta + (r_0 - r)v - \beta_0] = \Delta N[\beta + (r_0 - r)v + (R_c - r_0)v]$$

Recalling the definition of the two-phase formation volume factor u, this equation becomes

$$N[u - u_0] = \Delta N[u + (R_c - r_0)v] \qquad (3)$$

Equation 3 is known as the material balance equation for a constant volume reservoir with no initial gas cap and is a relation between the quantities of fluids in the reservoir and those produced. If the reservoir fluid characteristics are known or can be estimated the volume of oil remaining in the reservoir after a given production interval can be calculated, as is shown in the following example.

EXAMPLE. A constant volume reservoir with no gas cap had an original pressure of 2500 psia. The reservoir temperature is 150° F. 10^6 barrels of stock tank oil and 10^9 S.C.F. of gas were produced by the time the pressure had dropped to 2000 psia. Calculate the initial oil in place and oil remaining in the reservoir after this production interval. The reservoir fluid characteristics have the following values at the original reservoir pressure and at 2000 psia.

2500 psia	2000 psia
$r_0 = 825$ S.C.F./STB	$r = 680$ S.C.F./STB
$\beta_0 = 1.48$	$\beta = 1.39$
$u_0 = \beta_0 = 1.48$	$v = 0.0012$ Bbl/S.C.F.
	$u = \beta + (r_0 - r)v = 1.39 + (825 - 680)0.0012 = 1.56$

Solving the material balance equation 3 for N

$$N = \frac{\Delta N[u + (R_c - r_0)v]}{u - u_0} = \frac{10^6\left[1.56 + \left(\frac{10^9}{10^6} - 825\right)0.0012\right]}{1.56 - 1.48}$$

$$= 22.1 \times 10^6 \text{ stock tank barrels of oil initially in place}$$

Therefore, the oil remaining in the reservoir is $22.1 \times 10^6 - 1 \times 10^6 = 21.1 \times 10^6$ stock tank barrels.

Generalized Material Balance Equation for Reservoirs with an Initial Gas Cap and Water Encroachment. Constant volume reservoirs, such as the one considered in the previous section, are seldom met in actual practice. Usually the volume of the reservoir decreases as production progresses because formation water encroaches the reservoir. Furthermore, actual reservoirs often exist with the initial pressure below the bubble point so that a gas cap is present. Consequently, the material balance equation must be extended to include the initial gas cap and the effect of water encroachment.

All symbols used in the following derivation have their usual meaning. In addition it is necessary to define

m = ratio of *original* reservoir gas cap volume to the original reservoir oil volume.

W = cumulative water influx into reservoir in barrels.

w = cumulative water production in barrels.

$W - w$ = cumulative water encroachment. (This represents the decrease in reservoir volume.)

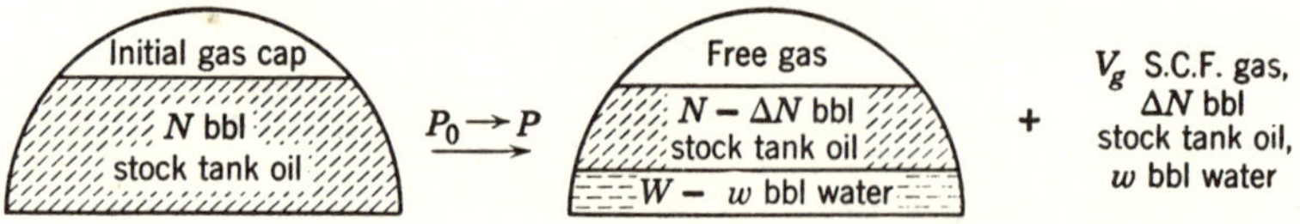

FIG. 94. Volume relationships for a production interval from a reservoir with an initial gas cap and water encroachment.

Figure 94 illustrates the process to be considered. A material balance on the gas yields

$$\begin{matrix}\text{S.C.F. of} & & \text{S.C.F. of gas} & & \text{S.C.F. of} & & \text{S.C.F. of} & & \text{S.C.F.} \\ \text{free gas in} & + & \text{in solution} & = & \text{free gas} & + & \text{gas remaining} & + & \text{of gas} \\ \text{initial gas cap} & & \text{originally} & & \text{in reservoir} & & \text{in solution} & & \text{produced}\end{matrix} \tag{4}$$

In terms of the reservoir fluid characteristics the terms in equation 4 become

$$\text{S.C.F. of gas in original gas cap} = \frac{mN\beta_0}{v_0}$$

$$\text{S.C.F. of gas originally in solution} = Nr_0$$

S.C.F. of gas in gas cap after an interval of production =

$$\frac{mN\beta_0 + [N\beta_0 - (N - \Delta N)\beta] - (W - w)}{v}$$

$$\text{S.C.F. of gas remaining in solution} = (N - \Delta N)r$$

$$\text{S.C.F. of gas produced} = V_g = R_c\Delta N$$

Substitution of these terms in equation 4 yields

$$\frac{mN\beta_0}{v_0} + Nr_0 = \frac{mN\beta_0 + [N\beta_0 - (\Delta - \Delta N)\beta] - (W - w)}{v} + (N - \Delta N)r + R_c\Delta N \tag{5}$$

Rearrangement and simplification of equation 5 gives

$$N\left[(u - u_0) + mu_0\left(\frac{v - v_0}{v_0}\right)\right] = \Delta N[u + (R_c - r_0)v] - W + w \tag{6}$$

Equation 6 is known as the generalized material balance equation. Values of ΔN, R_c, and w are usually available from production data. Values of u_0, u, r_0, v_0, and v are obtained from measured or estimated fluid characteristics. This leaves N, m, and W as unknowns and obviously two of these must be known in order to calculate the third.

Values of m and N are often available from electric log and core analysis data. If this is so, a value for the water influx W may be calculated as shown in the following example.

EXAMPLE. A field with a formation temperature of 150° F, an original pressure of 3000 psia, and 10^7 barrels of S.T. oil in place originally produced 10^6 barrels of 40° API oil, 5×10^4 bbl of water and 1.10×10^9 S.C.F. of 0.80 specific gravity gas by the time the pressure dropped to 2500 psia. What is the water influx and water encroachment if the gas cap volume was originally ¼ that of the original oil?

The values of the reservoir fluid characteristics are estimated by the methods of the previous chapter and found to be as follows:

$$r_0 = 1040 \text{ S.C.F./S.T. Bbl} \qquad \beta_0 = 1.58$$
$$r = 850 \text{ S.C.F./S.T. Bbl} \qquad \beta = 1.48$$
$$v_0 = 0.00080 \text{ Bbl/S.C.F.} \qquad u_0 = 1.58$$
$$v = 0.00092 \text{ Bbl/S.C.F.} \qquad u = 1.655$$

Solving equation 6 for W one has

$$W = \Delta N[u + (R_c - r_0)v] - N\left[(u - u_0) + mu_0\frac{(v - v_0)}{v_0}\right] + w$$

$$= 10^6\left[1.655 + \left(\frac{1.1 \times 10^9}{10^6} - 1040\right)0.00092\right] - 10^7\left[(1.655 - 1.58)\right.$$

$$\left. + \frac{1}{4} \times 1.58\left(\frac{0.00092 - 0.00080}{0.00080}\right)\right] + 5 \times 10^4 = 420{,}000 \text{ Bbl}$$

Therefore the water encroachment $= W - w = 420{,}000 - 50{,}000 = 370{,}000$ Bbl

It is evident that equation 3 is a special case of equation 6 since equation 6 reduces to equation 3 when m and $W - w$ are both equal to zero. Furthermore, it should be noted that in the derivation of equation 6 the water-formation volume factor and solubility of gas in water were not considered. This is in keeping with the results described in Chapter 6 since in most instances these effects are small and may be neglected.

Material Balance Equation for a Reservoir Producing Above the Saturation Pressure. When a reservoir exists at a pressure above its bubble point oil can be produced by expansion of the reservoir fluid as the pressure is reduced to the saturation pressure. This process is shown for a constant volume reservoir in Figure 95. The pressure declines from the original reservoir pressure to a pressure which is equal to or greater than the saturation pressure.

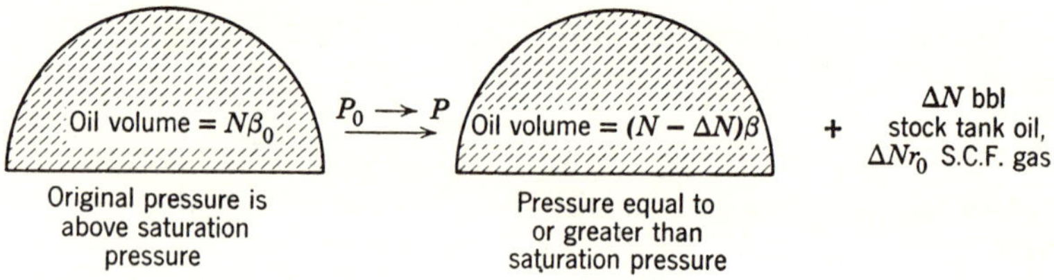

FIG. 95. Volume relationships for a reservoir producing above the saturation pressure.

Since the volume of the reservoir is assumed constant it is apparent that

$$N\beta_0 = (N - \Delta N)\beta \qquad \text{or} \qquad \frac{\Delta N}{N} = \frac{\beta - \beta_0}{\beta} \tag{7}$$

This equation relates the cumulative oil produced to the original oil in place and enables the fraction of oil produced by expansion of the reservoir fluid above its bubble point to be computed as shown in the following example.

EXAMPLE. What fraction of the oil in place can be produced from a reservoir on reduction of pressure from P_0 to P_s if $\beta_0 = 1.40$ and $\beta_s = 1.50$?
Application of equation 7 gives

$$\text{Fraction of oil produced} = \frac{\Delta N}{N} = \frac{\beta_s - \beta_0}{\beta_s} = \frac{1.50 - 1.40}{1.50} = 0.0666$$

Note that here again the generalized material balance equation as given by equation 6 would yield the same results under the conditions imposed. For a constant volume reservoir above its bubble point

$$u = \beta, \qquad u_0 = \beta_0, \qquad m = 0, \qquad R_c = r_0, \qquad W = 0,$$

and

$$w = 0$$

Substitution of these values in equation 6 yields equation 7.

Equation 7 may be readily extended to the case where the reservoir volume is decreased by water influx. If no water is produced it may be shown that equation 7 becomes

$$\frac{\Delta N}{N} = \frac{\beta - \beta_0}{\beta} + \frac{W}{\beta N} \tag{8}$$

The Reservoir Saturation Equation. The pore space within a porous medium which constitutes a petroleum reservoir may contain water, oil, and gas. The saturation of a given fluid is defined as the fraction or percentage of the *pore* space which is filled with that fluid. In many engineering calculations it is necessary to know the oil saturation in a reservoir after a production interval. It is possible to derive an equation for oil saturation in terms of the oil-formation volume factor and other measurable quantities when the following simplifying assumptions are made: (1) the oil remaining in the reservoir is *evenly* distributed throughout the pore space; (2) the saturation of connate water remains constant and no water is produced; (3) there is no water influx so that the pore volume of the reservoir is constant; (4) the oil is undersaturated so that no free gas is initially present in the pore space.

Consider a reservoir which initially contained N stock tank barrels of oil and which has produced ΔN stock tank barrels of oil. By definition the oil saturation after a production interval is given by:

$$\text{Oil saturation} = S_0 = \frac{\text{oil volume}}{\text{pore volume}} = \frac{(N - \Delta N)\beta}{\text{pore volume}} \tag{9}$$

If the constant connate water saturation is denoted by S_w, it is apparent that in the case where the initial gas saturation is zero the initial oil saturation is equal to $1 - S_w$. Therefore the pore volume is given by

$$\text{Pore volume} = \frac{N\beta_0}{1 - S_w} \tag{10}$$

Since it has been assumed that the pore volume of the reservoir is constant the pore volume given by equation 10 may be substituted in equation 9. Consequently

$$S_0 = \frac{(N - \Delta N)\beta}{\dfrac{N\beta_0}{1 - S_w}} = (1 - S_w)\left(1 - \frac{\Delta N}{N}\right)\frac{\beta}{\beta_0} \tag{11}$$

Example. A constant volume reservoir containing a constant connate water saturation of 0.20 and no initial free gas had an original oil-formation volume factor of 1.50. When 10% of the original oil was produced the value of the oil-formation volume factor decreased to 1.38. Calculate the oil saturation.

According to equation 11 the oil saturation is

$$S_0 = (1 - S_w)\left(1 - \frac{\Delta N}{N}\right)\frac{\beta}{\beta_0} = (1 - 0.20)(1 - 0.10)\,\frac{1.38}{1.50} = 0.662$$

Darcy's Law. Darcy's Law describes the flow of fluid through a porous medium. Consider a uniform porous medium which is *completely* saturated with a single homogeneous fluid. According to Darcy's Law the flow of fluid through the porous medium is directly proportional to the cross-sectional area and the pressure gradient, and inversely proportional to the fluid viscosity. In equation form Darcy's Law may be written

$$Q = -\frac{KA}{\mu}\frac{dP}{dL} \tag{12}$$

where Q is the volume of flow, A is the cross-sectional area, dP/dL is the pressure gradient, μ is the fluid viscosity and K is a proportionality constant known as the *permeability*. Equation 12 implies that the length L is measured in the direction of *decreasing* pressure so that the pressure gradient is negative and the minus sign in the equation is necessary. Certain other conditions are implied in equation 12. Not only must the porous medium be homogeneous and completely saturated with a homogeneous fluid but the flow must be non-turbulent, steady state, and isothermal. Furthermore, there must be no chemical interactions between the flowing fluid and the porous medium.

The unit of permeability is called a "darcy." A porous medium has a permeability of one darcy when a pressure gradient of one standard atmosphere per cm causes a flow of one cm^3 per sec of a one centipoise viscosity fluid through a porous medium which is one cm^2 in cross-sectional area.

Consider the system shown in Figure 96 in which the flow occurs through a constant cross-sectional area. For *liquid* flow through such

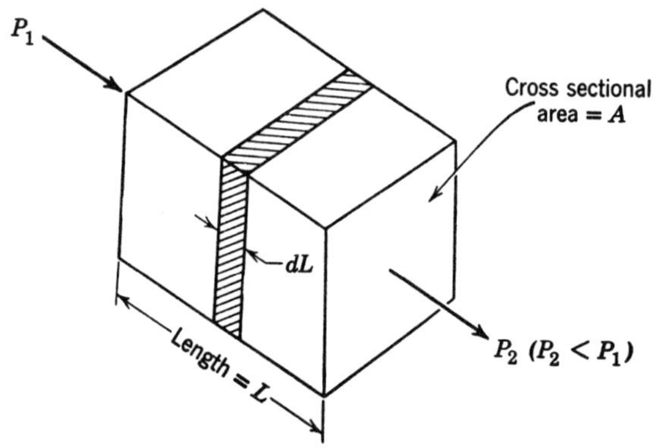

FIG. 96. Fluid flow in a linear system.

a linear system Q is not a function of pressure and equation 12 may be integrated directly as shown below.

$$\frac{Q}{A}\int_0^L dL = -\frac{K}{\mu}\int_P^{P_2} dP$$

$$Q = \frac{KA(P_1 - P_2)}{\mu L} \tag{13}$$

P_1 and P_2 denote the upstream and downstream pressures and L represents the length of the system.

EXAMPLE. A porous medium is 3 in. long and 1 in. in radius. 10 cm^3 of water ($\mu = 1$ c.p.) flows through the porous medium in 100 sec when the pressure drop is 30.4 in. of mercury. Calculate the permeability in darcys.

It is essential that each term in equation 13 be expressed in the proper units. If K is to be in darcys, Q must be in cm^3 per sec, μ in c.p., L in cm, A in cm^2, and $P_1 - P_2$ in standard atm. Solving equation 13 for K gives

$$K = \frac{Q\mu L}{A(P_1 - P_2)} = \frac{\frac{10}{100} \times 1 \times (3 \times 2.54)}{(\pi 2.54^2) \times 30.4/29.9} = 0.0369 \text{ darcys}$$

In the integration of equation 12 to give equation 13 it was assumed that Q was not a function of pressure. However, when a compressible fluid such as a gas is flowing this assumption is not valid. As the gas flows through the porous medium from a high pressure to a low pressure it expands as the pressure decreases. Consequently for a compressible fluid Q must be measured at the mean pressure of the system, that is, at a pressure equal to $(P_1 + P_2)/2$. If Boyle's Law applies to the gas it is evident that

$$P_m Q_m = PQ$$

where P_m is the mean pressure, Q_m is the quantity of flow measured at mean pressure and P and Q are the pressure and corresponding quantity of flow at any other pressure. Equation 12 may be written

$$\frac{Q}{A}\int_0^L dL = -\frac{K}{\mu}\int_{P_1}^{P_2} dP$$

and since $Q = \dfrac{P_m Q_m}{P}$ this equation becomes

$$\frac{P_m Q_m}{A}\int_0^L dL = -\frac{K}{\mu}\int_{P_1}^{P_2} P\,dP$$

On integration, one obtains

$$\frac{P_m Q_m L}{A} = \frac{K}{\mu}\left(\frac{P_1^2}{2} - \frac{P_2^2}{2}\right)$$

or

$$\frac{P_1 + P_2}{2}\,\frac{Q_m L}{A} = \frac{K}{2\mu}(P_1 - P_2)(P_1 + P_2)$$

or

$$Q_m = \frac{KA}{\mu L}(P_1 - P_2) \tag{14}$$

This equation has the same form as equation 13 except the quantity of flow is measured at the mean pressure. It is evident that gas flow and liquid flow in a linear system may be described by the same equation provided the quantity of flow is always measured at the mean pressure.

EXAMPLE. A core is 3 in. long and 2 cm in diameter. When the upstream pressure was 29.4 psia and the downstream pressure was 14.7 psia, 10 cc of air ($\mu = 0.018$ c.p.), measured at the downstream pressure, flowed per second. Calculate the permeability of the core in darcys.

$P_1 = 2$ atm and $P_2 = 1$ atm. Therefore $P_m = (1 + 2)/2 = 1.5$ atm. Since $P_m Q_m = P_2 Q_2$,

$$Q_m = \frac{1 \times 10}{1.5} = 6.67 \text{ cc per sec}$$

Consequently,

$$K = \frac{Q_m \mu L}{A(P_1 - P_2)} = \frac{6.67 \times 0.018 \times (3 \times 2.54)}{\pi \times 1} = 0.291 \text{ darcy}$$

Equation 12 may be used to derive an equation that approximates the flow from a surrounding reservoir into a producing well. Consider the system shown in Figure 97 where x_w and x_e represent the well

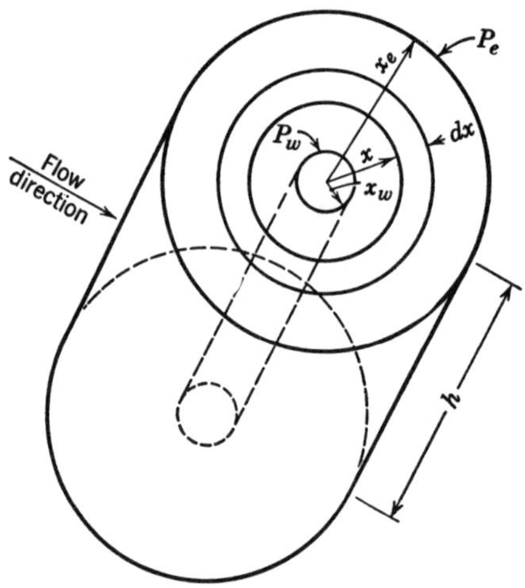

FIG. 97. Radial system which approximates the flow from a reservoir into a producing well.

radius and the external radius of the system, respectively, P_w and P_e represent the pressures at the well bore and at the external radius, and h represents the height of the system (thickness of the producing formation). Consider a cylindrical shell of radius x and thickness dx. If equation 12 is expressed in cylindrical coordinates, one has

$$Q = \frac{KA}{\mu}\left(\frac{dP}{dx}\right) \tag{15}$$

Since the area A equals $2\pi hx$, equation 15 becomes

$$Q = \frac{2\pi Khx}{\mu}\left(\frac{dP}{dx}\right)$$

Integrating between limits for a *non-compressible* fluid gives

$$Q\int_{x_w}^{x_e}\frac{dx}{x} = \frac{2\pi Kh}{\mu}\int_{P_w}^{P_e} dP$$

or

$$Q = \frac{2\pi Kh(P_e - P_w)}{\mu \ln \dfrac{x_e}{x_w}} \tag{16}$$

It can be shown that the equation for radial flow of a *compressible* fluid is

$$Q_m = \frac{2\pi Kh(P_e - P_w)}{\mu \ln \dfrac{x_e}{x_w}} \tag{17}$$

where Q_m is the volume of fluid flowing per second measured at the mean pressure of $(P_e + P_w)/2$. Here again it is evident that both liquid flow and gas flow may be calculated using the same equation if the rate of flow is measured at the mean pressure.

The system of units used to define the unit of permeability is not well suited for engineering calculations. For practical applications the following system of units is apparently more convenient:

Q or Q_m are measured in barrels per day
h is measured in feet
x_w and x_e are measured in feet
$P_e - P_w$ are measured in psi
K is measured in darcys
μ is measured in centipoise

It can be shown that with this system of units equations 16 and 17 become

$$Q = \frac{7.07Kh(P_e - P_w)}{\mu \ln \dfrac{x_e}{x_w}} \tag{18}$$

and

$$Q_m = \frac{7.07Kh(P_e - P_w)}{\mu \ln \dfrac{x_e}{x_w}} \tag{19}$$

The following example illustrates the use of these equations.

EXAMPLE. For a radial system x_e is 527 feet and x_w is 0.375 feet. If the permeability of the producing formation is 0.15 darcys and the thickness is 10 feet, calculate the quantity of 2 c.p. oil that will flow under a pressure drop of 500 psi.

Substitution into equation 18 gives

$$Q = \frac{7.07Kh(P_e - P_w)}{\mu \ln \frac{x_e}{x_w}} = \frac{7.07 \times 0.15 \times 10 \times 500}{2 \times \ln \frac{527}{0.375}} = 365.8 \text{ barrels/day}$$

In the previous discussion of Darcy's Law only homogeneous fluid flow was considered. If the porous medium is only partially saturated with a fluid the permeability of the porous medium to the fluid will be less than the permeability obtained when the medium is 100% saturated. The permeability at less than 100% saturation is known as the *effective* permeability. Values of the effective permeability range between 0 and K where K represents the permeability at 100% saturation. The symbols K_o, K_g, and K_w are used to represent the effective permeabilities to oil, gas, and water, respectively.

Effective permeability is used in Darcy's Law in place of the permeability at 100% saturation when two or more fluids are present in a porous medium. Thus, if oil partially saturates a porous medium, Darcy's Law may be written

$$Q_o = \frac{K_o A \Delta P}{\mu_o L}$$

Similarly for a porous medium partially saturated with gas the quantity of gas flowing is

$$Q_g = \frac{K_g A \Delta P}{\mu_g L}$$

and for water

$$Q_w = \frac{K_w A \Delta P}{\mu_w L}$$

The *relative* permeability is defined as the ratio

$$\text{Relative permeability} = \frac{\text{effective permeability at a given saturation}}{\text{permeability at 100\% saturation}}$$

The symbols K_o/K, K_g/K, and K_w/K are used to represent the relative permeabilities to oil, gas, and water, respectively. Obviously, relative permeability values range between 0 and 1. It has been found that, for a given porous medium, the *relative permeability is a function of saturation.* Consider a system in which both oil and gas are flowing simultaneously. For a system of this kind Figure 98 represents a typical plot of the two relative permeabilities as a function of saturation.

These relative permeability curves have several characteristics which appear to apply quite generally to most two-fluid systems. As shown in Figure 98, the oil relative permeability is zero below a finite oil saturation known as the equilibrium oil saturation and then increases with increasing oil saturation until it reaches a value of one

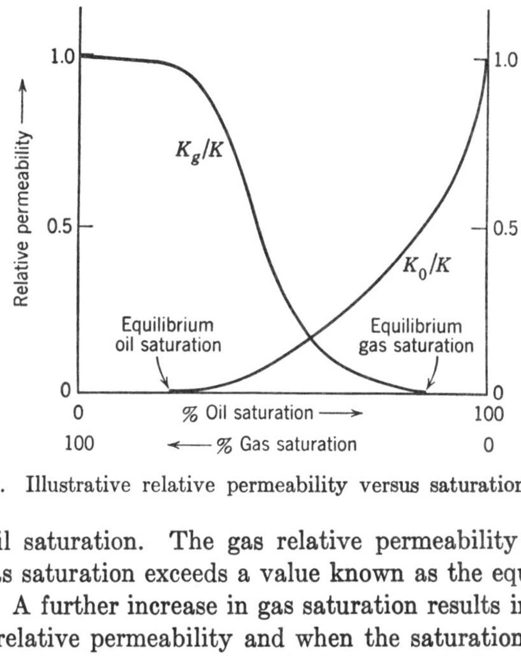

FIG. 98. Illustrative relative permeability versus saturation curves.

at 100% oil saturation. The gas relative permeability is also zero until the gas saturation exceeds a value known as the equilibrium gas saturation. A further increase in gas saturation results in an increase in the gas relative permeability and when the saturation corresponds to the equilibrium oil saturation K_g/K is essentially equal to one.

These characteristics of the relative permeability curves are intimately related to the fluid distribution in the porous medium. For the system under consideration, oil is the wetting phase, that is, it clings to the grain surface of the porous medium while the gas occupies the open channels, as shown in Figure 99. For very low oil saturations the oil forms "pendular" rings around the grain contact points. These rings do not touch one another so that for all practical purposes the liquid is immobile. Consequently $K_o/K = 0$. The presence of these pendular rings of oil have but little effect on the flow of gas so that small oil saturations do not decrease K_g/K appreciably. As the saturation of oil increases the pendular rings grow in size. At the equilibrium oil saturation the pendular rings touch one another and the oil begins to flow. It flows with difficulty but the gas in the still open channels flows with relative ease. As the oil saturation continues to increase K_o/K increases and K_g/K decreases. Eventually the oil saturation becomes high enough so that the gas can no longer flow in a continuous stream but breaks up into discrete bubbles. At the equilibrium gas saturation all the gas exists as immobile discrete bubbles and K_g/K is zero. Moreover, the immobile gas bubbles existing in the channels hinder the flow of oil and K_o/K is considerably less than one at these small gas saturations. Although the flow mechanism

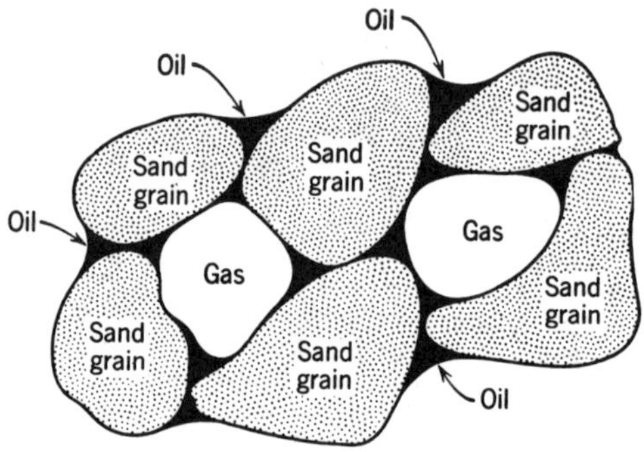

FIG. 99. Distribution of gas and oil in a porous medium.

described above may be over-simplified, it does explain the characteristics of the relative permeability versus saturation curves in a simple and reasonable manner.

In the case of oil and gas flow considered above, the oil phase is always the wetting fluid. When two liquids flow simultaneously in a porous medium either one can be the wetting phase depending on the nature of system. Thus, in the case of oil and water flow, either the oil or the water can be the wetting liquid depending on whether the solid composing the porous medium is preferentially water wet or oil wet.

A useful way of expressing relative permeability data as a function of saturation is by means of the relative permeability ratio. The gas-oil relative permeability ratio is defined as

$$\frac{K_g/K}{K_o/K} = \frac{K_g}{K_o}$$

and is obviously a function of saturation for a given porous medium. Figure 100 illustrates the functional relationship between K_g/K_o and oil saturation. K_g/K_o is infinite at the equilibrium oil saturation and then

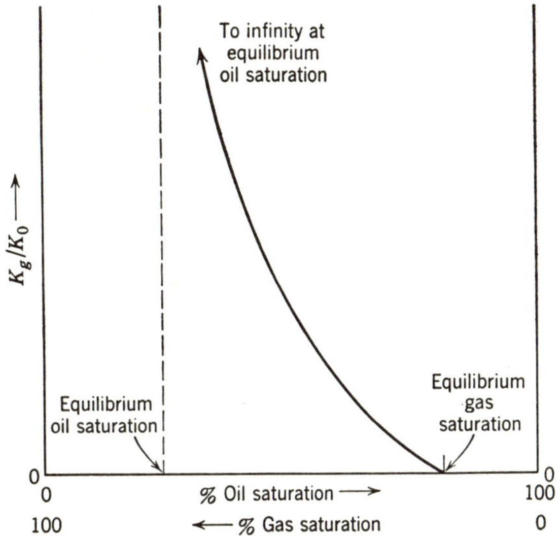

FIG. 100. Typical gas-oil relative permeability ratio versus saturation curve for a system containing oil and gas only.

decreases with increasing oil saturation until it becomes equal to zero at the equilibrium gas saturation.

The Producing Gas-Oil Ratio Equation. This equation, which is also known as the instantaneous gas-oil ratio equation, is of fundamental importance in reservoir analysis. To derive this equation consider a radial flow system shown in Figure 97. Consider a dx thickness within the system at radius x. If the pressure gradient at this radius is dP/dx, Darcy's Law in differential form for the simultaneous flow of oil and gas *in the reservoir* may be written

$$Q_o = \frac{7.07K_o hx}{\mu_o}\frac{dP}{dx}$$

$$Q_g = \frac{7.07K_g hx}{\mu_g}\frac{dP}{dx}$$

where Q_o = volume of oil flowing in barrels per day in the reservoir
Q_g = volume of gas flowing in barrels per day in the reservoir
K_o = effective oil permeability
K_g = effective gas permeability
μ_0 = viscosity of oil
μ_g = viscosity of gas
h = thickness of the radial system

If the volume of oil is to be measured in stock tank barrels, then

$$\text{Stock tank barrels of oil per day} = q_o = \frac{Q_o}{\beta}$$

Similarly, if the volume of gas is to be measured in standard cubic feet then it is apparent that

$$\text{S.C.F. of gas per day} = q_g = \frac{Q_g}{v} + rq_o$$

since each stock tank barrel of oil will liberate r S.C.F. of gas when the pressure is reduced to atmospheric. Consequently, by definition

$$\text{Producing gas-oil ratio} = R = \frac{q_g}{q_o} = \frac{Q_g\beta}{Q_o v} + r$$

$$= \frac{7.07K_g hx/\mu_g\ (dP/dx) \times \beta}{7.07K_o hx/\mu_o\ (dP/dx) \times v} + r$$

$$= \frac{K_g \mu_o \beta}{K_o \mu_g v} + r \tag{20}$$

if it is assumed that the pressure gradient is the same in both the oil phase and the gas phase.

The quantities in equation 20 must be evaluated at the pressure existent at the radius x. This, in general, is not known but, if the well is shut in the average reservoir pressure can be measured. In using equation 20 β, v, r, μ_o, and μ_g are usually evaluated at the average reservoir pressure. This is equivalent to assuming that production occurs at zero pressure differential, that is, with a zero pressure drawdown.

Another assumption that is implicit in this derivation is that the oil and gas are distributed uniformly and are flowing according to the relative permeability concept previously described.

Example. The gas-oil relative permeability ratio for a producing formation is 0.0162. The oil-gas viscosity ratio is 42.4. Values of β, r, and v are known to be 1.37, 580 S.C.F./S.T. Bbl and 0.001122 barrels/S.C.F. respectively at the average reservoir pressure. Calculate the producing gas-oil ratio.

Using equation 20

$$R = \frac{K_g \mu_o \beta}{K_o \mu_g v} + r = 0.0162 \times 42.4 \times \frac{1.37}{0.001122} + 580 = 1420 \text{ S.C.F. of gas/stock tank barrel of oil}$$

The producing gas-oil ratio and cumulative gas-oil ratio defined in the previous sections on reservoir material balances should be clearly distinguished. The latter was defined as the ratio of *all* the gas produced to *all* the oil produced during the production history of a reservoir. The producing gas-oil ratio on the other hand is measured over a short time interval. The relationship between the producing gas-oil ratio and the cumulative gas-oil ratio is illustrated in Figure 101. In Figure 101 the producing gas-oil ratio is plotted as a function of

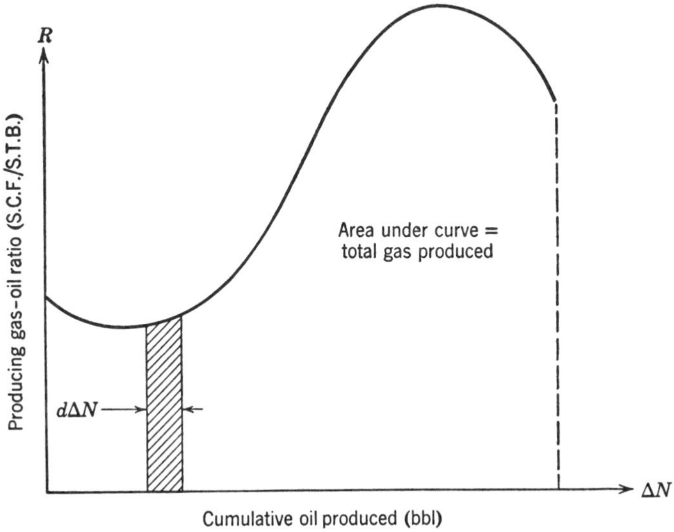

FIG. 101. Relationship between producing gas-oil ratio and total gas produced.

oil produced. For a small production interval represented by $d\Delta N$ the gas produced is

$$R d\Delta N$$

Consequently, the cumulative gas produced is given by the integral

$$V_g = \int_0^{\Delta N} R d\Delta N \tag{21}$$

which represents the area under the R versus ΔN curve from 0 to ΔN. By definition R_c is

$$R_c = \frac{V_g}{\Delta N} = \frac{\text{area under } R \text{ versus } \Delta N \text{ curve}}{\Delta N}$$

Prediction of Oil Reservoir Behavior. By combining the material balance equation, the reservoir saturation equation, the producing gas-oil ratio equation, and the equation relating the cumulative gas-oil ratio to the producing gas-oil ratio, it is possible to compute the producing gas-oil ratio and the oil produced with declining reservoir pressure. In order to simplify the problem a constant volume reservoir with no initial gas cap will be considered although it is to be understood that more complex reservoirs may also be treated. Furthermore, it will be assumed that there is no segregation of oil and gas within the reservoir as production proceeds and that no water is produced.

In the reservoir under consideration the energy available for expulsion of oil and gas comes entirely from the evolution of solution gas on pressure reduction. Consequently, this type of reservoir is designated as a solution gas drive reservoir to distinguish it from those whose recovery mechanisms involve energy from the expansion of a gas cap (gas expansion reservoirs) or from the encroachment of water (water drive reservoirs). The behavior of a solution gas drive reservoir may be predicted if the following data are available: (1) the original reservoir pressure and temperature; (2) values of r, β, and v as a function of pressure; (3) values of the reservoir fluid viscosities as a function of pressure at reservoir temperature; (4) the constant water saturation (S_w); (5) values of K_g/K_o as a function of saturation; and (6) the number of barrels of stock tank oil originally in reservoir (N). The computations are carried out stepwise as shown below.

Step 1. At a pressure a few hundred pounds below P_0 choose a value of ΔN and solve the material balance equation for a constant volume reservoir with no initial gas cap (equation 3) for the number of standard cubic feet of gas produced.

$$\text{S.C.F. gas produced} = V_g = \Delta N R_c = \frac{N(u - u_0) - \Delta N(u - r_0 v)}{v}$$

Step 2. Using the same value of ΔN calculate S_0 using equation 11

$$S_0 = (1 - S_w)\left(1 - \frac{\Delta N}{N}\right)\frac{\beta}{\beta_0}$$

Step 3. The value of S_0 computed in step 2 determines the relative permeability ratio at that saturation. Using this value of K_g/K_o compute the producing gas-oil ratio using equation 20

$$R = \frac{K_g \mu_o \beta}{K_o \mu_g v} + r$$

Step 4. The value of R computed in step 3 represents a point on the R versus ΔN curve. According to equation 21 the area under the R versus ΔN curve is equal to V_g. A numerical value for the area may be calculated if a linear relationship between R and ΔN is assumed *within a given production interval.* Consequently, at the first pressure below P_0 the area is given by

$$V_g = \left(\frac{r_0 + R}{2}\right) \Delta N$$

and at subsequent pressures

$$V_g = \sum \frac{R_i + R_f}{2} (\Delta N_f - \Delta N_i)$$

where R_i and R_f are the initial and final producing gas-oil ratios in the production interval represented by $(\Delta N_f - \Delta N_i)$.

Step 5. The value of V_g obtained in step 4 should agree with the value for V_g computed in step 1. If the two values do not agree the computations are repeated at another value of ΔN.

When the true value of ΔN has been determined by this trial-and-error process a second pressure a few hundred pounds below the first is chosen and the process repeated. By a series of computations of this type the future behavior of the reservoir may be predicted.

EXAMPLE. A constant volume reservoir without an initial gas cap has an original pressure equal to the saturation pressure of 2500 psia and a temperature of 180° F. The connate water saturation is 20% and no water is produced. There are 56 × 10^6 stock tank barrels of oil originally in place. If the reservoir fluid characteristics are those given in Table 12, and the values of K_g/K_0 as a function of oil saturation are those given in Figure 102, calculate R and ΔN as a function of pressure.

Table 12. Reservoir Fluid Characteristics

Pressure	β	r S.C.F./S.T.B.	$v \times 10^3$ Bbl/S.C.F.	μ_o c.p.	μ_g c.p.	u $\beta + (r_0 - r)v$
2500	1.498	721	1.048	0.488	0.0170	1.498
2300	1.463	669	1.155	0.539	0.0166	1.523
2100	1.429	617	1.280	0.595	0.0162	1.562
1900	1.395	565	1.440	0.658	0.0158	1.620
1700	1.361	513	1.634	0.726	0.0154	1.701
1500	1.327	461	1.884	0.802	0.0150	1.817
1300	1.292	409	2.206	0.887	0.0146	1.967
1100	1.258	357	2.654	0 981	0.0142	2.251
900	1.224	305	3.300	1.085	0.0138	2.597
700	1.190	253	4.315	1.199	0.0134	3.209
500	1.156	201	6.163	1.324	0.0130	4.361
300	1.121	149	10.469	1.464	0.0126	7.109
100	1.087	97	32.032	1.617	0.0122	21.075

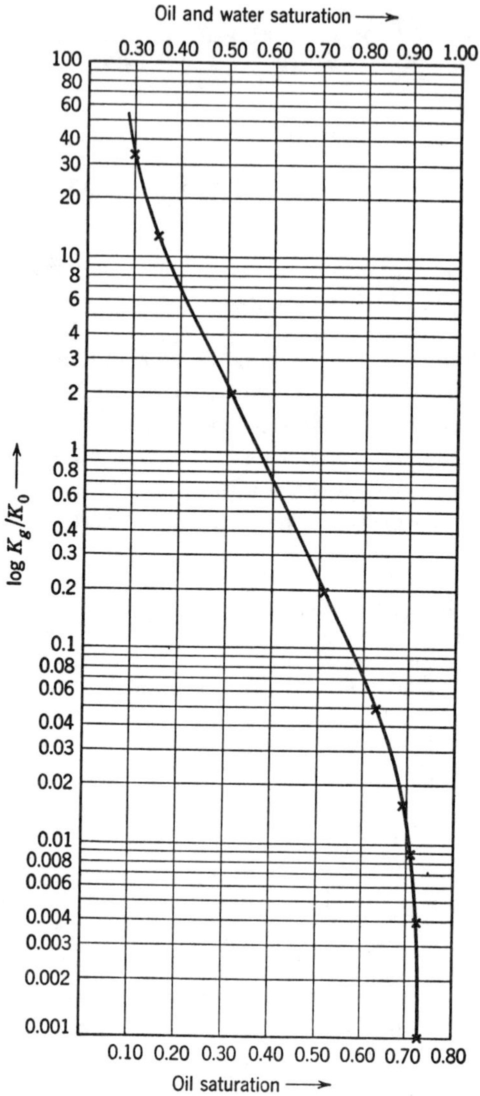

FIG. 102. Logarithm of K_g/K_0 as a function of oil saturation (or oil + water saturation). The water saturation is constant and equal to 0.20.

The computations at 2300 and 2100 psia are presented below. At 2300 psia assume a value of ΔN equal to $0.01675N$. Substitution in equation 3 leads to

$$V_g = \Delta N R_c = \frac{N(u - u_0) - \Delta N(u - r_0 v)}{v}$$

$$= \frac{N(1.523 - 1.498) - 0.01675N(1.523 - 721 \times 0.001155)}{0.001155}$$

$$= 11.64N$$

Using the same value of ΔN calculate S_0 as required by step 2 using equation 11

$$S_0 = (1 - S_w)\left(1 - \frac{\Delta N}{N}\right)\frac{\beta}{\beta_0} = (1 - 0.2)(1 - 0.01675)\frac{1.463}{1.498} = 0.7682$$

At this oil saturation the relative permeability ratio is zero, as shown in Figure 101. Consequently, the producing gas-oil ratio is given by

$$R = \frac{K_g \mu_o \beta}{K_o \mu_g v} + r = r = 669 \text{ S.C.F./Bbl}$$

The area under the R versus ΔN curve is equal to

$$V_g = \frac{r_0 + R}{2}\Delta N = \frac{721 + 669}{2}(0.01675N) = 11.64N$$

Since this value of V_g agrees with that obtained above by the material balance equation, the value of ΔN assumed at 2300 psia is correct. Consequently, by the time the reservoir pressure drops to 2300 psia the producing gas-oil ratio is 669 S.C.F./Bbl and the volume of stock tank oil produced is

$$0.01675N = 0.01675 \times 56 \times 10^6 = 9.38 \times 10^5 \text{ S.T.B.}$$

It is understood that if these two values of V_g did *not* agree a new value of ΔN would have been chosen and the computation repeated. In practice it is customary to judiciously choose two or three values of ΔN and calculate the two values of V_g obtained for each value of ΔN. Then the values of V_g calculated by material balance are plotted versus ΔN and the values of V_g obtained by computing the area under the R versus ΔN curve are plotted versus ΔN. Obviously, the point of intersection of these two curves occurs at the true value of ΔN.

To continue the calculation assume a value of ΔN equal to $0.0425N$ at 2100 psia. V_g by material balance is given by

$$V_g = \Delta N R_c = \frac{N(1.562 - 1.498) - 0.0425N(1.562 - 721 \times 0.00128)}{0.00128}$$
$$= 28.78N$$

The oil saturation is computed to be

$$S_0 = (1 - 0.2)(1 - 0.0425)\frac{1.429}{1.498} = 0.731$$

At this value of S_0 the relative permeability ratio is found to be 0.001 from Figure 101. Accordingly, the producing gas-oil ratio is

$$R = 0.001 \times \frac{0.595}{0.0162} \times \frac{1.429}{0.00128} + 617 = 658 \text{ S.C.F./Bbl}$$

The area under the R versus ΔN curve is equal to the area from $\Delta N = 0$ to $\Delta N = 0.01675N$, as determined above, plus the area from $\Delta N = 0.01675N$ to $\Delta N = 0.0425N$. The total area is given by

$$V_g = 11.64N + \frac{669 + 658}{2}(0.0425N - 0.01675N) = 28.73N$$

Since this value again agrees sufficiently well with that calculated by the material balance equation the producing gas-oil ratio at 2100 psia is 658 S.C.F./Bbl and the number of stock tank barrels of oil produced is

$$0.0425 \times 56 \times 10^6 = 2.38 \times 10^6 \text{ S.T.B.}$$

The calculations are continued in this manner at pressure increments of 200 psi. The results obtained are summarized in Table 13 and are presented graphically in Figure 103.

Table 13. Calculation of R and ΔN as a Function of Pressure

Pressure psia	R S.C.F./Bbl	$\Delta N/N$	ΔN S.T. Barrels
2500	721	0	0
2300	669	0.01675	0.938×10^6
2100	658	0.0425	2.38×10^6
1900	1330	0.0704	3.94×10^6
1700	1940	0.0935	5.24×10^6
1500	2430	0.1143	6.40×10^6
1300	3080	0.1312	7.35×10^6
1100	3830	0.1527	8.55×10^6
900	4400	0.1663	9.31×10^6
700	4820	0.1850	10.36×10^6
500	4780	0.2001	11.20×10^6
300	4200	0.2177	12.19×10^6
100	2160	0.2440	13.66×10^6

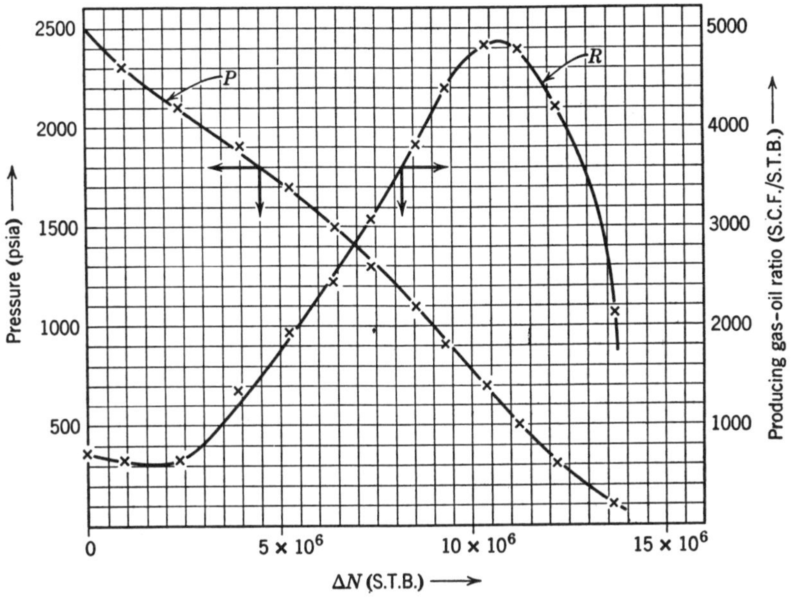

FIG. 103. Plot of pressure and producing gas-oil ratio as a function of stock tank barrels of oil produced.

The observant reader will notice that the trial-and-error method of solution outlined above is not necessary and ΔN may be solved for directly when the gas saturation is *less* than the equilibrium gas saturation. As long as this condition is maintained gas cannot flow and the producing gas-oil ratio is equal to the gas in solution. For example, at the first pressure below P_0 it is apparent that

$$\frac{N(u - u_0) - \Delta N(u_0 - r_0 v)}{v} = \frac{r_0 + r}{2} \Delta N$$

and this equation may be solved for ΔN directly.

REFERENCES

Calhoun, J. C., *Fundamentals of Reservoir Engineering,* University of Oklahoma Press, Norman, Okla. (1953).

Muskat, M., *Physical Principles of Oil Production,* McGraw-Hill Book Co., New York (1949).

Pirson, S. J., *Elements of Oil Reservoir Engineering,* McGraw-Hill Book Co., New York (1950).

Standing, M. B., *Volumetric and Phase Behavior of Oil Field Hydrocarbon Systems,* Reinhold Publishing Corp., New York (1952).

PROBLEMS

1. Show that equation 6 (Chapter 7) follows from equation 5.

2. A reservoir originally contained 42×10^6 bbl S.T. oil. The gas cap volumn was 1630 acre-ft. At the original pressure of 2000 psia, r_0 was 530 S.C.F./bbl, β_0 was 1.37 and v_0 was 0.00110 bbl/S.C.F. After 1.70×10^6 bbl S.T. oil, 1273×10^6 S.C.F. of gas and 90,000 bbl of water had been produced, the pressure was 1800 psia. At this pressure r was 481 S.C.F./bbl, β was 1.34, and v was 0.00122 bbl/S.C.F. Find W, the water influx. *Answer:* W = 290,000 bbl.

3. The original data from a reservoir are: P_0 = 2500 psia, r_0 = 710 S.C.F./bbl, β_0 = 1.435, v_0 = 0.00092 bbl/S.C.F., $N = 72 \times 10^6$ bbl S.T. oil. When the pressure was 2200 psia the data were r = 620, β = 1.388, v = 0.00110, $\Delta N = 3.51 \times 10^6$ bbl S.T. oil, R_c = 1260 S.C.F./bbl, w = 0. (*a*) Assume the reservoir volume is constant and find the apparent value of m. *Answer:* m = 0.178; (*b*) Assume m = 0 and find the apparent value of W.

4. Using β, r, v, u, N, ΔN, etc., write expressions for the following: (*a*) Volume of original oil in reservoir. *Answer:* $N\beta_0$; (*b*) Standard cubic feet of gas dissolved per barrel of *reservoir* oil; (*c*) Volume occupied in the reservoir in barrels by the gas which comes out of solution from one barrel of stock tank oil; (*d*) Cumulative gas-oil ratio in S.C.F./S.T.B. for production above the saturation pressure.

5. Show that the conversion factor in equations 18 and 19 is 7.07.

6. A ¾ in. diameter plug is cut from a core and trimmed to a length of 2⅛ in. After drying it is found to weigh 32.30 grams. When 100% saturated with water it weighed 35.53 grams. A 37° A.P.I. oil (μ = 4 c.p.) and water (μ = 1 c.p.) were flowed through the core and the following data taken

Core Weight, grams	Δt, sec	Vol H_2O, cc	Vol Oil, cc	ΔP, psi
35.53	200	20.7	0	12
35.49	250	24.1	0	14
35.47	250	30.2	0.01	20
35.43	250	22.2	0.20	20
35.34	250	12.1	1.08	20
35.30	300	11.0	2.74	25
35.26	300	6.45	3.71	25
35.22	300	2.26	5.32	25
35.16	300	0.64	7.93	25
35.11	300	0.05	11.4	25
35.09	300	0	13.6	25

Calculate K_o/K and K_w/K and plot as a function of water saturation.

7. The following data were taken on a small field

Well No.	Producing Gas-Oil Ratio, S.C.F./Bbl	Daily Production of Oil, Bbl
1	1320	215
2	1729	170
3	1370	205
4	1219	234
5	1440	195

What is the average producing gas-oil ratio for the field? *Answer:* 1398 S.C.F./Bbl.

8. For a field the following data have been accumulated for the producing gas-oil ratio as a function of total oil produced

Producing G.O.R., S.C.F./Bbl	Total Oil Produced, Bbl
730	0
715	2×10^3
660	11.2×10^3
1020	19.6×10^3
1642	29.6×10^3
2416	36.7×10^3
2650	40.0×10^3

What would be the cumulative G.O.R. after (*a*) 29.6×10^3 barrels of oil were produced? (*b*) after 40.0×10^3 barrels of oil were produced?

9. After producing 15% of the original oil from a constant volume reservoir containing no initial free gas the following data are available

R = 280 S.C.F/Bbl $\qquad r$ = 410 S.C.F./STB
β = 1.30 $\quad \beta_0$ = 1.45 $\qquad \mu_0/\mu_g$ = 58.2
v = 0.001795 Bbl/S.C.F. $\qquad$ Constant water saturation = 30%

Calculate the gas-oil relative permeability ratio and the oil saturation. *Answer:* $K_g/K_o = 0.0567$.

10. Using the data presented in Table 12 and Figure 102, calculate the values of R and ΔN that have been summarized in Table 13.

APPENDIX A

SYMBOLS AND ABBREVIATIONS

A	Cross-sectional area
AMW	Apparent molecular weight
° API	API Gravity $= \frac{141.5}{SG} - 131.5$
a	Van der Waals constant
α	Coefficient of thermal expansion
BPP	Bubble-point pressure
Btu	British thermal unit
β	Oil-formation volume factor
β_0	Original oil-formation volume factor at initial reservoir temperature and pressure
β_s	Oil-formation volume factor at saturation pressure
β_w	Water-formation volume factor
Bbl	Barrels
b	Van der Waals constant
C	Coefficient of compression
° C	Centigrade degree ° C = $5/9$ (° F − 32)
c.p.	Centipoise
γ	Shrinkage factor
D	Density
DPP	Dew-point pressure
ΔH_m	Molar heat of vaporization
D_l	Liquid density
D_v	Vapor density
° F	Fahrenheit degree
h	Height (thickness)
° K	Kelvin degree
K	Permeability
K_o	Effective permeability of oil
K_g	Effective permeability of gas

K_w	Effective permeability of water
L	Length
ln	Natural logarithm
log	Logarithm to base 10
MW	Molecular weight
m	Ratio of original gas cap volume to original oil volume
N	Stock tank barrels originally in reservoir
n	Number of moles
P°	Vapor pressure
P	Pressure
P_c	Critical pressure
$\underline{P_c}$	Pseudo-critical pressure
P_0	Original pressure
P_R	Reduced pressure
P_s	Saturation pressure
Q	Quantity of flow per second measured in reservoir
q	Quantity of flow per second at stock tank conditions (60° F and 14.7 psia)
R	Gas constant or producing gas-oil ratio
° R	Rankine degree
R_c	Cumulative gas-oil ratio
r	Gas solubility (S.C.F./S.T.B.)
S.C.F.	Standard cubic feet
S.G.	Specific gravity
S.T.B.	Stock tank barrel
S_0	Saturation of oil
S_w	Saturation of water
T	Temperature
T_c	Critical temperature
$\underline{T_c}$	Pseudo-critical temperature
T_0	Reservoir temperature
T_R	Reduced temperature
t	Time
u	Two-phase formation volume factor
μ	viscosity
V	Volume
V_g	S.C.F. of gas produced
v	Gas-formation volume factor
W	Barrels of water influx
Wt	Weight
w	Barrels of water produced
x	Mole fraction in liquid or radius
y	Mole fraction in vapor
z	Mole fraction in entire system
Z	Compressibility factor

APPENDIX B

PHYSICAL AND THERMODYNAMIC PROPERTIES OF METHANE

Temperature °F	Vapor Pressure psia	Density of Liquid lb/cu ft	Density of Vapor lb/cu ft	Heat of Vaporization Btu/lb
−280	4.4	27.5		229.4
−270	8.0	27.0		225.3
−260	14.7	26.5	0.1123	221.3
−250	23.1	26.0	0.169	217.2
−240	33.1	25.5	0.234	213.0
−230	48.1	24.9	0.330	208.7
−220	64.5	24.4	0.435	204.0
−210	89.9	23.8	0.585	199.0
−200	120.8	23.2	0.748	193.4
−190	155.4	22.5	0.975	187.1
−180	197.9	21.7	1.24	179.7
−170	241.7	21.0	1.56	171.2
−160	293.0	20.2	1.93	161.4
−150	338.8	19.5	2.31	149.5
−140	412.3	18.5	2.93	134.4
−130	539.0	16.1	4.93	112.6

PHYSICAL AND THERMODYNAMIC PROPERTIES OF ETHANE

Temperature °F	Vapor Pressure psia	Density of Liquid lb/cu ft	Density of Vapor lb/cu ft	Heat of Vaporization Btu/lb
−120	19.0	33.7	0.163	197.9
−110	24.7	33.3	0.209	195.7
−100	32.2	32.8	0.267	193.5
−90	41.1	32.3	0.335	191.0
−80	51.5	31.9	0.414	188.4
−70	64.4	31.5	0.514	185.4
−60	78.1	31.0	0.616	182.2
−50	95.5	30.5	0.724	178.6
−40	113	30.0	0.836	174.8
−30	137.3	29.6	1.00	170.6
−20	161.9	29.1	1.26	166.1
−10	190.5	28.6	1.49	161.0
0	222.3	28.0	1.77	155.4
10	257.8	27.3	2.08	149.0
20	296.7	26.7	2.44	141.8
30	338.0	26.1	2.82	133.7
40	389.4	25.3	3.34	124.5
50	442.3	24.4	3.89	113.9
60	503.0	23.4	4.70	101.0
70	568.3	22.1	5.76	84.6
80	641.3	19.9	7.70	61.6
89.4	708.5	13.73	13.73	0.000

PHYSICAL AND THERMODYNAMIC PROPERTIES OF PROPANE

Temperature °F	Vapor Pressure psia	Density of Liquid lb/cu ft	Density of Vapor lb/cu ft	Heat of Vaporization Btu/lb
−70	7.37	37.40	0.0775	189.5
−60	9.72	37.00	0.111	187.0
−50	12.6	36.60	0.129	184.5
−40	16.2	36.19	0.163	181.5
−30	20.3	35.78	0.203	179.0
−20	25.4	35.37	0.250	176.0
−10	31.4	34.96	0.307	173.5
0	38.2	34.54	0.369	170.5
10	46.0	34.12	0.441	168.0
20	55.5	33.67	0.526	165.0
30	66.3	33.20	0.625	162.0
40	78.0	32.73	0.730	159.0
50	91.8	32.24	0.847	156.0
60	107.1	31.65	0.990	153.0
70	124.0	31.24	1.13	149.5
80	142.8	30.70	1.30	146.0
90	164.0	30.15	1.49	142.5
100	187.0	29.58	1.69	138.5
110	212.0	28.96	1.92	134.0
120	240.0	28.30	2.18	129.0

PHYSICAL AND THERMODYNAMIC PROPERTIES OF NORMAL BUTANE

Temperature °F	Vapor Pressure psia	Density of Liquid lb/cu ft	Density of Vapor lb/cu ft	Heat of Vaporization Btu/lb
0	7.3	38.59	0.090	170.5
10	9.2	38.24	0.112	168.5
20	11.6	37.89	0.138	167.0
30	14.4	37.54	0.169	165.5
40	17.7	37.19	0.205	163.5
50	21.6	36.82	0.246	161.5
60	26.3	36.44	0.294	159.5
70	31.6	36.06	0.347	157.5
80	37.6	35.65	0.407	155.0
90	44.5	35.24	0.476	152.0
100	52.2	34.84	0.552	149.5
110	60.8	34.41	0.633	147.0
120	70.8	33.96	0.725	143.5
130	81.4	33.49	0.826	140.5
140	92.6	32.98	0.934	137.5

PHYSICAL AND THERMODYNAMIC PROPERTIES OF NORMAL PENTANE

Temperature °F	Vapor Pressure psia	Density of Liquid lb/cu ft	Density of Vapor lb/cu ft	Heat of Vaporization Btu/lb
60	7.0	39.30	.....	.....
100	15.6	37.96	0.197	153.8
110	19.0	37.61	0.236	150.7
120	22.5	37.25	0.274	147.9
130	26.6	36.88	0.322	146.5
140	31.0	36.50	0.374	144.0
150	36.6	36.12	0.440	141.2
160	42.3	35.73	0.508	138.7
170	48.9	35.34	0.585	136.2
180	54.1	34.93	0.669	133.3
190	57.4	34.51	0.766	129.9
200	67.5	34.08	0.871	127.5
210	82.3	33.64	0.980	126.2
220	94.6	33.20	1.13	122.6
230	106.1	32.75	1.26	120.0
240	119.7	32.26	1.43	116.7
250	133.4	31.76	1.60	114.1

APPENDIX C

TABLE FOR CONVERTING API GRAVITY TO SPECIFIC GRAVITY

Degrees API	Specific Gravity	Degrees API	Specific Gravity	Degrees API	Specific Gravity	Degrees API	Specific Gravity
0	1.076	26	0.898	52	0.771	78	0.675
1	1.068	27	0.893	53	0.767	79	0.672
2	1.06	28	0.887	54	0.763	80	0.669
3	1.052	29	0.882	55	0.759	81	0.666
4	1.043	30	0.876	56	0.755	82	0.663
5	1.037	31	0.871	57	0.75	83	0.659
6	1.028	32	0.865	58	0.747	84	0.656
7	1.02	33	0.860	59	0.743	85	0.654
8	1.014	34	0.855	60	0.739	86	0.65
9	1.007	35	0.85	61	0.735	87	0.647
10	1.00	36	0.845	62	0.731	88	0.645
11	0.993	37	0.840	63	0.728	89	0.642
12	0.986	38	0.835	64	0.724	90	0.639
13	0.979	39	0.83	65	0.72	91	0.636
14	0.973	40	0.825	66	0.716	92	0.633
15	0.966	41	0.82	67	0.713	93	0.63
16	0.959	42	0.816	68	0.709	94	0.628
17	0.953	43	0.811	69	0.706	95	0.625
18	0.946	44	0.806	70	0.702	96	0.622
19	0.94	45	0.802	71	0.699	97	0.619
20	0.934	46	0.797	72	0.695	98	0.617
21	0.928	47	0.793	73	0.692	99	0.614
22	0.922	48	0.788	74	0.689	100	0.611
23	0.916	49	0.784	75	0.685	101	0.608
24	0.910	50	0.779	76	0.682	102	0.606
25	0.904	51	0.775	77	0.679	103	0.604

TABLE FOR CONVERTING API GRAVITY TO SPECIFIC GRAVITY *(Continued)*

Degrees API	Specific Gravity	Degrees API	Specific Gravity	Degrees API	Specific Gravity	Degrees API	Specific Gravity
104	0.602	118	0.567	132	0.537	146	0.51
105	0.598	119	0.565	133	0.535	147	0.508
106	0.596	120	0.563	134	0.533	148	0.506
107	0.594	121	0.56	135	0.531	149	0.504
108	0.591	122	0.558	136	0.529	150	0.502
109	0.589	123	0.556	137	0.527	151	0.501
110	0.586	124	0.554	138	0.525	152	0.499
111	0.584	125	0.552	139	0.523	153	0.497
112	0.581	126	0.55	140	0.521	154	0.496
113	0.579	127	0.547	141	0.519	155	0.494
114	0.576	128	0.545	142	0.517	156	0.493
115	0.574	129	0.543	143	0.515	157	0.49
116	0.572	130	0.541	144	0.514	158	0.489
117	0.570	131	0.539	145	0.512	159	0.487

INDEX

Absolute temperature, 17, 30, 41, 45
Acetylene hydrocarbons, 8
API gravity, 108
Apparent molecular weight, 22, 87, 134
Aromatic hydrocarbons, 9 ff.
Asphalt, 12
Avogadro's Law, 18

Benzene hydrocarbons, 9
Boyle's Law, 16 ff., 43, 163
Bubble point, 53, 58, 81, 85, 88, 94 ff.

Charles' Law, 17 ff.
Clausius-Clapeyron equation, 44 ff., 90
Coefficient, of compression, 39, 121, 140
  of thermal expansion, 39
Composition, in two-phase region, 83 ff., 88 ff., 92 ff.
  of connate water, 135
  of natural gas, 12
  of petroleum, 10 ff.
Compressibility factor, 28 ff., 103, 147 ff.
Connate water, 135
Cracking, 13
Cricondenbar, 60
Cricondentherm, 60
Critical locus, 65
Critical pressure, 29, 49, 59
Critical temperature, 29, 49, 54, 59
Cumulative gas-oil ratio, 155, 170 ff.
Cycloparaffins, 8

Dalton's Law of partial pressures, 26 ff., 81, 84, 88
Darcy's Law, 161 ff.
Density of a perfect gas, 19, 23
Density-temperature diagram, 55 ff.
Dew point, 53, 58, 86, 94 ff.
Differential liberation of gas, 106, 112, 145
Diolefin hydrocarbons, 7

Effective permeability, 166
Equilibrium constants, 90 ff.
Extensive properties, 48

Flash liberation of gas, 106, 112, 145

Gas constants, 18 ff., 45
Gas-formation volume factor, 101, 103 ff.
Gas gravity, 22 ff.
Gas gravity measurement, 23 ff.
Gas mixtures, 19 ff.
Gas-oil ratio, 155, 169, 170
Gasoline, 13 ff.
Gas solubility, 102, 105 ff., 135 ff., 143 ff.

General gas law, 18
Geneva system, 5, 7, 8
Gibb's phase rule, 75 ff.
Graham's Law of diffusion, 25

Heat of vaporization, 44, 46
Henry's Law, 98, 108, 136

Ideal solutions, 80 ff.
Instantaneous gas-oil ratio, 169 ff.
Intensive properties, 48, 75
Interstitial water, 135
Isomerism, 4, 7, 10

Kelvin temperature scale, 17, 45

Law of Rectilinear Diameters, 55 ff.

Material balance equation, 155 ff., 171
Mole fraction, 20, 67
Multicomponent systems, 72 ff.

Naphthene hydrocarbons, 8 ff.
Natural gas, 12, 32
Natural gasoline, 13
Nomenclature of hydrocarbons, 4 ff., 7, 8, 9
Non-ideal solutions, 90 ff.

Oil-formation volume factor, 102, 110 ff., 143 ff.
Olefin hydrocarbons, 6

Paraffin hydrocarbons, 2 ff., 13
Perfect gas laws, 16 ff.
Permeability, 162
Prediction of reservoir behavior, 171 ff.
Pressure-composition diagram, 65 ff., 82, 87
Pressure-temperature diagrams, 49 ff., 59 ff., 63 ff., 73 ff.
Pressure-volume diagrams, 52 ff., 57 ff., 72 ff.
Producing gas-oil ratio equation, 169 ff.
Products of petroleum, 13 ff.
Pseudo-critical pressure and temperature, 30, 33, 65, 104, 134
Pseudo-reduced pressure and temperature, 32, 104, 134

Radial flow equation, 164 ff.
Rankine temperature scale, 17, 45
Raoult's Law, 81, 83, 88
Reduced pressure and temperature, 29
Relative permeability, 166 ff.
Relative permeability ratio, 168
Reservoir saturation equation, 160 ff.
Retrograde phenomena, 60 ff., 75

Saturation equation, 160 ff.
Saturation pressure, 73, 105, 110, 121, 138, 145, 159 ff.
Shrinkage factor, 110
Shrinkage of oil, 110, 114
Single-component systems, 49 ff.
Solution gas drive reservoir, 171 ff.
Solutions, 79 ff.
Specific gravity of gases, 22 ff.
Standard conditions, 18, 102, 103

Tar, 12
Temperature-composition diagram, 71 ff.
Triple point, 50
Trouton's Rule, 46, 90
Two-component systems, 57 ff.
Two-phase formation volume factor, 102, 123 ff.

Unsaturated hydrocarbons, 6 ff.

Van der Waals' equation, 27 ff.
Vapor pressure, 39 ff., 49, 80 ff.
Vapor pressure measurement, 41 ff.
Viscosity, of gas, 102, 128 ff., 151 ff.
  of oil, 102, 126 ff., 149 ff.
  of water, 141 ff.

Water-formation volume factor, 138 ff.
Water encroachment, 157
Water influx, 157

Z factor, 28 ff., 103, 147 ff.